T0316746

Radical Reporting

Most people dread writing reports; they also dread reading reports. What they don't realize is that the techniques that make writing more readable make it more powerful. This is especially relevant for professionals in areas such as audit, risk, compliance, and information security.

This small volume provides the tools and techniques needed to improve reports. It does so through addressing crucial concepts all too often overlooked in the familiar rush to perform tasks, complete projects, and meet deadlines.

These concepts – the role of culture in communication; the link between logic and language; the importance of organizing thoughts before writing; and how to achieve clarity – may seem academic or theoretical. They're not. Unless writers understand their own thoughts, actions, and objectives, they cannot hope to communicate them at all – let alone clearly.

This second edition develops these points with additional material on critical thinking, as well as the use of AI in reporting.

Sara I. James is an internationally recognized expert in internal audit communications. Through her business, Getting Words to Work®, she delivers tailored report-writing and other training to internal audit, risk, compliance, and information security teams worldwide (www.saraijames.com). With over 30 years' academic, teaching, writing, publishing, and corporate experience in North America, Europe, the Middle East, and the Far East, she brings a wealth of varied yet specialized expertise to clients and audiences. As a member of the Chartered Institute of Internal Auditors (UK and Ireland) Technical Guidance Working Group, Sara has produced many governance, risk, and internal audit advisory pieces. She has also produced articles on clear writing for industry journals and spoken at numerous audit conferences and heads of internal audit forums. In 2022, she was named one of Richard Chambers' Internal Audit Beacon award-winners. Finally, Sara has been fortunate to live, study, and work in different countries, including the US, the UK, France, and the then-Soviet Union, and speaks five languages. These experiences and skills give her invaluable insight into cross-cultural communication and global organizations. She is also regularly asked to contribute scholarly essays, articles, and conference papers on language and literature.

Security, Audit and Leadership Series

Series Editor: *Dan Swanson, Dan Swanson and Associates, Ltd., Winnipeg, Manitoba, Canada.*

The *Security, Audit and Leadership Series* publishes leading-edge books on critical subjects facing security and audit executives as well as business leaders. Key topics addressed include Leadership, Cybersecurity, Security Leadership, Privacy, Strategic Risk Management, Auditing IT, Audit Management and Leadership

Global Audit Leadership: A Practical Approach to Leading a Global Internal Audit (GIA) Function in a Constantly Changing Internal and External Landscape
Audley L. Bell

Construction Audit: Building a Solid Foundation
Denise Cicchella

Continuous Auditing with AI in the Public Sector
Lourens Erasmus and Sezer Bozkus Kahyaoglu

Ironwill 360° Leadership: Moving Forward: Unlock Twelve Emerging Trends for Forward Thinking Leaders
Douglas P. Pflug

The CISO Playbook
Andres Andreu

Leveraging Blockchain Technology: Governance, Risk, Compliance, Security, and Benevolent Use Cases
Shaun Aghili

The Closing of the Auditor's Mind?: How to Reverse the Erosion of Trust, Virtue, and Wisdom in Internal Auditing
David J. O'Regan

Radical Reporting: Writing Better Audit, Risk, Compliance, and Information Security Reports (Second Edition)
Sara I. James

For more information about this series, please visit: www.routledge.com/Internal-Audit-and-IT-Audit/book-series/CRCINTAUDITA

Radical Reporting

Writing Better Audit, Risk, Compliance, and Information Security Reports

Second Edition

Sara I. James

CRC Press
Taylor & Francis Group
Boca Raton London New York

CRC Press is an imprint of the
Taylor & Francis Group, an **informa** business

Second edition published 2025
by CRC Press
2385 NW Executive Center Drive, Suite 320, Boca Raton FL 33431

and by CRC Press
4 Park Square, Milton Park, Abingdon, Oxon, OX14 4RN

CRC Press is an imprint of Taylor & Francis Group, LLC

© 2025 Sara I. James

First edition published by CRC Press 2022

ISBN: 978-1-032-72754-7 (hbk)
ISBN: 978-1-032-72753-0 (pbk)
ISBN: 978-1-003-42236-5 (ebk)

DOI: 10.1201/9781003422365

Typeset in Sabon
by Newgen Publishing UK

To the late Les Greenblatt – cousin, teacher, mentor, friend.
And to Chessh – first and last and always.

Contents

Acknowledgments

This section alone risks seeing me break one of my own rules – not to write too much. But how else can I thank everyone who has helped me with this book, an exciting project and one I hope will change how people work.

From the beginning, this has been an international endeavor, though one that started close to home, when my partner, John Chesshire, put me in touch with CRC Press series editor Dan Swanson. Dan, together with Gabriella Williams and her colleagues at Taylor & Francis Group, made writing the book a far more enjoyable process than I ever would have imagined.

Thanks to those who reviewed my initial proposal – David Deegan, Liz Sandwith, and Meredythe Spence – for their time, encouragement, and insights. Along with the subject-matter experts, Martin Cutts and John Chesshire, they made comments and suggestions that vastly refined and improved my original thoughts. Thanks, too, to Dr. Sumner Braund, Professor Robert F. Cook, and Professor David Lee Rubin for their practical help and advice with the mechanics of editing, indexing, and graphics. The manuscript would have been a sorry-looking creature indeed without you.

The content and graphics wouldn't have existed without contributions from colleagues and friends across the world. From Amsterdam to Yerevan, people with incredible experience, wisdom, and advice generously agreed to provide graphics, templates, and their own insights during interviews. Thanks to Pierre André, Rupert Bamberger, Ola Bello, Jane Bettany, Paul Breach, Jeffrey W. Brown, Rachel Browne, Mark Carawan, Ara Chalabyan, Richard Chambers, Vincent Chardot, David Dart, David Deegan, David Hill, Rinus de Hooge, Nino Karazanashvili, Philipp Kratzer, Alee Marschke, Margaret McCaig, Louise McKay, Fran Meegan, Marisa Melliou, Rupert Morris, Duyen Nguyen, Virpi Oinonen, Scott Petersen, Alex Psarras, Nannette Ripmeester, Alexander Rühle, Steven Sanders, Liz Sandwith, Dan Shoemaker, Emma Smith, Komitas Stepanyan, Alex Tytkovskyi, Shawn Von Hagen, Malcolm Zack, and all those who contributed anonymously.

My final words on this page are for those who gave me my first words: my parents, Tony and Alice Tortora. Thank you for passing onto me your lifelong love of language, reading, writing, and learning.

Author

Sara I. James is an internationally recognized expert in internal audit communications. Through her business, Getting Words to Work®, she delivers tailored report-writing and other training to internal audit, risk, compliance, and information security teams worldwide (www.saraijames.com). With over 30 years' academic, teaching, writing, publishing, and corporate experience in North America, Europe, the Middle East, and the Far East, she brings a wealth of varied yet specialized expertise to clients and audiences.

As a member of the Chartered Institute of Internal Auditors (UK and Ireland) Technical Guidance Working Group, Sara has produced many governance, risk, and internal audit advisory pieces. She has also produced articles on clear writing for industry journals and spoken at numerous audit conferences and heads of internal audit forums. In 2022, she was named one of Richard Chambers' Internal Audit Beacon award-winners.

Finally, Sara has been fortunate to live, study, and work in different countries, including the US, the UK, France, and the then-Soviet Union, and speaks five languages. These experiences and skills give her invaluable insight into cross-cultural communication and global organizations. She is also regularly asked to contribute scholarly essays, articles, and conference papers on language and literature.

Introduction

AND SO IT BEGINS

If you have bought this book, or even borrowed it, you must be in desperate straits. After all, reporting is painful – and anything radical doubly so.

You may see reporting, and individual reports, as a problem to solve. True enough, they tend to linger on most office workers' to-do lists. Many people admit to loathing reports – both writing and reading them. Some may claim to love writing them, to find the process easy: "The words just flow," they say. I suspect their readers don't always love reading the result quite so much.

However, whether you love or loathe writing or reading reports, you probably see far too many bad ones. Bad reports do not exist in isolation, of course. They are often the product of individual assumptions and habits, team and organizational culture, and over-engineered processes.

Bad reports are a problem because they are usually painful to write and read. So the people and teams producing them suffer unnecessarily, as do readers. This brings us to an even bigger problem than that of individual suffering.

If we write bad reports, people won't want to read them. Those who force themselves to do so may find they cannot understand them. Worse, they may believe they understand them – but they don't.

All these problems have real-world consequences. If we don't recognize and solve the problems, then our reports will be at best ineffective and at worst damaging. By this, we mean the very purpose of reports – to provide information that helps readers improve things – is thwarted. This in turn has serious repercussions for entire organizations, especially in times of crisis.

Will this book help you solve at least some of these problems? I hope so. It depends on your expectations, really. You should reconsider your purchase if any of the following apply:

- You want a list of rigid rules, dos, and don'ts so that you can simply copy and paste "prefabricated"[1] words and phrases onto a blank screen.

- Similar to the above, you think there is only one way to write a report, and this book will tell you what it is.
- You think your colleagues, team members, and even line managers are all idiots. You therefore hope to use this book as ammunition next time they have the gall to write something in a style other than yours.
- You have an important exam coming up next week and think this book will help you pass it. (Not a chance. Most exams simply require you to regurgitate definitions and phrases; this book wants you to think critically.)
- You don't want to think critically. Ever.

If you've lasted this long, you may have an inkling that writing a good report is not a speedy or superficial matter. It requires serious reflection. It also requires accepting some ambiguity, as well as having a splash of tolerance for others. (This may be particularly difficult for my target audience: audit, risk, compliance, and IT professionals. But I say this from a place of love.)

Still with me? Good – you clearly recognize there is a problem and want to solve it. But to do this, we need to go to the very roots of writing and thus reporting. This is why the book is called *Radical Reporting*.

Let's begin with some definitions.[2]

Report:

- An account given of a particular matter, especially in the form of an official document, after thorough investigation or consideration by an appointed person or body.
- A spoken or written description of an event or situation, especially one intended for publication or broadcasting in the media.

Radical:

- (Especially of change or action) relating to or affecting the fundamental nature of something; far-reaching or thorough.
- Forming an inherent or fundamental part of the nature of someone or something.
- (Of surgery or medical treatment) thorough and intended to be completely curative.

So how will this book help you improve your reports, making them truly radical? Read on.

First, it will encourage and help you to think deeply and originally, I hope, about writing. Why do you write anything, let alone reports? What do you expect from the process and the product? Do you think about what others expect?

Writers must understand the purpose of any writing, including writing reports. Speaking of purpose, are reports even necessary? Surprisingly, not always. For example, the Global Internal Audit Standards of the Institute of Internal Auditors refers to communicating results, but doesn't dictate using a report to do so.

However, since we must communicate results, most of us still find it better to record what we need to say in some permanent form. We could use cave painting or metaphysical poetry, but for now, reports seem to do the trick. So how do we make reports fulfill a real function, rather than becoming an activity performed for its own sake?

This is an uncomfortable question. If you are in a team dedicated to reporting, querying the very purpose of reports may seem obvious. "We are the regulatory risk reporting team – we know why we write reports!" But do you? I've seen regulatory risk reports that inform, warn, instruct, shift blame, ask questions, and much more, all in the same document. This is not necessarily a good thing – as a reader, I never once felt certain of the purpose of the report and therefore my role in reading it. Did the writers intend all that? Or did they, faced with a deadline and some internally sensitive content, start filling pages in a panic? I'm not being cynical – merely observant. Even if we are required to produce a report that does all I've described, it's best to do so understanding *why*.

We must also understand the poor readers who have to work through reports, hunting for meaning, advice, and recommendations. After all, they're not reading them for fun, but because they must. They do so usually in good faith, believing that their efforts will lead to something useful. Writers create a product; readers consume it. This is, as Steven Pressfield rightly says, "above all, a transaction. The reader donates his time and attention, which are supremely valuable commodities. In return, you the writer must give him something worthy of his gift to you."[3]

Understanding readers means not only knowing their job titles. It also means recognizing how culture – from team and organizational culture, to regional and national culture – affects communications. Individual response also plays a part; you can find yourself addressing a team and receiving starkly different responses from individuals within it. Some may be open to what you say, others defensive or even combative. Cultures may play a part, but so will personalities and feelings.

Don't run away at the mention of feelings! The emotional aspects of writing, reviewing, and reading play a major role in this book. For our writing to be effective, we must recognize and anticipate our and others' reactions to writing.

You, too, will have your own fears, preconceptions, and workload, all of which affect how you write. This is why the first, and possibly the most important chapter in this book, addresses the role of culture in all communication. It's not something that applies only to global companies or multicultural teams. We all have our own approaches, assumptions, and reactions, especially under stress or pressure.

Speaking of pressure, you may say, "I don't have time for this. I know why I have to write this report and why people should read it. I just need to sit down and write the stupid thing, now!" And you may well punch out 10, 20, 30, even 50 pages by a deadline. But are you sure anyone will read it?

This first step of serious, critical reflection about *why* and *for whom* we write is essential. I agree it takes time that we often feel we do not have to spare. Yet without it, anything we write will not attract, retain, or persuade readers. And that would be a false economy. As Abraham Lincoln supposedly said, "If I had five minutes to chop down a tree, I'd spend the first three sharpening my axe."

This book's focus is reports. However, to produce better reports, we have to go to the root (radical, remember?). This means laying the groundwork for critical thinking and a conscious appreciation of the complexity and nuance of language. When you find yourself writing on autopilot, the reader will definitely experience that boredom and fatigue described earlier. That's not good for them, for you, or for what the report is trying to achieve.

The opposite of autopilot is conscious effort: critical thinking, certainly, but also the time, discipline, and focused self-doubt that implies. It's essential for any successful communication.

Does that mean writing is hard? Well, it depends on what you mean. Spewing black marks onto a screen, or scribbling by hand without engaging your brain – unfortunately, that's all too easy. The advent of word processing hasn't made people worse writers. It's simply multiplied the amount of bad writing we read, just as it has multiplied the amount of (much rarer) good writing we read.[4]

Writing *well*, on the other hand, is hard, as the great William Zinsser said: "A clear sentence is no accident."[5] It requires thought, discipline, even selflessness. And who wants to put that much effort into anything?

If you want people to read what you write, *you* should want to, or at least recognize the need for effort. Writers as different as Byron and Hemingway have agreed: "Easy writing makes hard reading." There must be effort on one side of the transaction between writer and reader – and if the writer doesn't take the trouble, why should the reader? Most people will say the problem is time. They will acknowledge that good, clear, useful writing takes time. ("I'm sorry to have sent you such a long letter; I didn't have time to write a short one.") This prompts the question: time for what? Writing? Correcting? Arguing with colleagues and managers about the text? None of these.

What this book will encourage you to do first is to sit still and think. You may say you don't have time, but again, that's a false economy. Failing to think clearly about what you have to say, and why, means you will never convey it well.

The real reason, though, is that it's uncomfortable. No one wants to be alone with their thoughts. Yet if you don't master this crucial step, how can you clarify and then articulate those thoughts, so that someone else can translate them into action?

If you think skipping the initial, unavoidable stage of reflection is OK, ask yourself why you're reading this book. Are your reports good enough? Are they even readable?

If they aren't good enough or readable enough, perhaps it's because you're wasting time and energy on the wrong things. This may mean readers waste time and energy trying to understand what you've agonized to write.

The second way in which this book will help you improve your reports is by offering concrete tools and techniques. As we've said, virtually everyone nowadays would say they have too much to do and too little time in which to do it. That's one reason I despise the common advice to draft a document and then put it away overnight, or even for a week, before re-reading it afresh. What planet do people who say that live on? Can I visit? If you are in a position to leave drafts unbothered for days at a time, then you probably don't need this book.

For those of us living on this planet, however, this book offers practical, relevant advice and resources that should help you work more efficiently and effectively. These are aids to identify quickly and clearly the focus of your writing and to diagnose common problems. There are no magic tricks, or rigid lists of dos and don'ts (comforting though it may be to outsource our judgment), though. Instead, this book provides aids not only for your own reflections, insights, drafting, and reviewing, but also for sharing your newfound knowledge.

Of course, this book aims to help you improve all written communication in the workplace: annual, periodic, or engagement plans; terms of reference; memos; emails; project updates; business cases; and so on. In this respect, it is not unique.

There are many resources on reporting, and any sound book, course, or trainer will give you many of the same principles on writing clearly. This book offers numerous resources for you to broaden and deepen your knowledge of good writing, and good reporting. Other books in this series will complement this book by helping you expand your communication skills.

However, you still need something different – something that gives you even more. The "more" is insight specifically into some of the hardest reports to read and write: audit, risk, and compliance reports, including those from IT and cyber specialists. Their subject matter may appear dry and technical, but at every stage, human beings and their feelings are involved. To help you produce better reports, no matter how technical, each chapter will feature advice from a range of practitioners in audit, risk, compliance, and IT security. Throughout the book, you will see insights and suggestions from people who set the standards in their organizations and professions.

Readers are human, too. When they complain reports are boring, it's often because the subjects they convey are dry. Often, however, what passes for boredom is actually fatigue. It's that creeping weariness, sometimes accompanied by a looming headache, that signals sustained yet pointless mental effort.

This is what happens when reports – or any documents – are badly written. Most people write badly, which turns reports into one of the most reliable horrors of corporate life. As we said above, reports aren't isolated documents – they're the result of bad reporting processes, practices, and products, all of which I'll discuss in later chapters.

All writers must acknowledge these and other uncomfortable truths. Above all, to communicate well in any medium, writers must expend time, effort, and discipline: three things many of us feel we lack. This is why it's important to have relevant, practical tools and techniques we can use to improve our writing and communicate action.

As we said earlier, this book won't be the only one trying to help you out of bad habits. It is, however, the only one that combines that element with an in-depth knowledge of some of the toughest areas to produce reports in. These areas – audit, risk, and compliance, especially with an IT or cyber slant – see people trying to document what has happened or what they've done. The dictionary definitions support this. However, it's only half of the equation – and an ineffectual half, if you don't consider the other.

The other half of the equation is why people document what they do. You may say it's primarily to cover yourself, and few would disagree. One thing this book does not do is sugarcoat the reality of office politics, corporate standards, and team culture. All these things and more conspire to militate against good, clear communication.

However, many people want to produce something worthwhile, which will leave readers clear about what they should do and why, whether that's improving operations or risk mitigation or project delivery. Both groups, which overlap, will find what they need here.

The third way this book will help you is by boosting your confidence. I'm convinced that many potentially good writers who nonetheless produce stilted, wordy, vague reports do so because somewhere along the way they have lost confidence. Maybe a teacher mocked them; maybe a colleague red-penned an early draft into oblivion. Maybe, through years of seeing and being encouraged to produce bad writing, they've become convinced that their original way – clear, simple, tell-it-like-it-is – was somehow foolish, unsophisticated, or unprofessional.

That's all wrong. If you recognize yourself in what I've just described, then consider this book your de-programming. Some of the invitations to reflect may be uncomfortable. You may have to examine some of the assumptions, values, and habits you've acquired over the years. Are they useful? Were they ever? What will you have to do to change them, and how will you feel about that? (See, we're back to feelings.)

However, I'm convinced that when this book asks you to pause and reflect, even five minutes will produce sound insights. You can then use these insights to identify your strengths and focus on which one or two weaknesses you need to address, to produce better reports more quickly and confidently.

I've already mentioned the practical advice and resources this book offers. There are also exercises. All of these aim to build your skills and confidence. Now, given the pace of change in the 21st century, pandemic-inspired changes in working customs, and people's limited time, we have to choose those tools that simplify and streamline.

It is therefore up to you how much you do, when, and how. Even if you only dip into the book initially, you should find suggestions that are immediately relevant to your work. You can (and should) always revisit it later, perhaps doing some of the exercises or selecting resources for further reading. You're in charge!

Once you've built up your own confidence, and mastered the techniques that work for you, you can share what you've learned with your colleagues. Whether it's having more probing conversations with reviewers, reviewing peers' work more constructively, or coaching colleagues, your progress doesn't end with the last page. You'll know that this book has done its job when you realize that your writing skills are a work in progress – forever.

I strongly believe that only professional writers can advise on writing. This is because non-professionals – experienced subject-matter experts, for example – may be excellent at what they do, but that doesn't mean they can also communicate it well. We can all put black marks on a page, but few of us are experts in language or writing. After all, simply having a body doesn't make one a doctor.

Many people also tend to see their own experiences and preferences as a universal truth. When they try to translate this into coaching or training, they risk narrowing people's understanding of what's possible, and worse, passing on bad habits or myths. You will have seen this in the workplace. Communication and report-writing in particular are subject to a lot of debate, not least because people have strong views.

What complicates matters further is the debates about the language used: vocabulary, grammar, punctuation, usage, tone. Often this is because ineffective and inefficient use of language obscures the content. However, it's also because writing and reviewing are deeply personal actions. Language is a medium to express perception, understanding, judgment, emotion, and choice. It goes to the heart of our beliefs and even our sense of self. Consider that dry IT report. If you criticize the language, you criticize the writer's personal style. If you criticize grammar and punctuation, you criticize their education, background, and upbringing. If you criticize how they have structured and worded the report, you may be criticizing the way they see the world.

Writing and reviewing reports is a minefield. That's why this book will give you a manageable range of tools and techniques for every stage of writing your report (or any other document). They are practical, time-tested, and apply even when you're on version 37, it's past midnight, and the draft was due last week. What's more, if you follow the advice in this book, you should never find yourself in that situation again.

In the following chapters, you will discover insights, approaches, and resources to help you re-build your reporting skills from the inside out. There will be case studies, exercises, recommended further reading, and end-of-chapter summaries and activities to help you improve day by day. However, you will get more out of them once we have laid the foundations of our work, by examining in detail the purpose of communication, and the underlying assumptions we bring to it. The first few chapters, focusing on these topics, are thorough if not lengthy. However, just as a sturdy house needs solid foundations, so too does effective communication.

Going back to our principle of radical reporting, the root of good reporting is understanding communication – what, why, who, when, and where? The first topic we will cover will address how every act of communication takes place within a cultural context: national, sector- or profession-specific, and organizational, in a section entitled "Words in the mind." Building on this, we will then look at the mishaps and pitfalls inherent in intra-organizational communication. Many people will have experienced the pain of trying to persuade senior managers that their risk management is flawed. How many have experienced even more pain discussing the same topic within their own teams?

The book then moves on to what you may have expected (and can easily find in other books): writing clearly. This section is called "Words on the page." Making the case for clear thinking leading to clear writing, we'll then show how to achieve it. Grammar, punctuation, and usage play a part, but this book will demystify long-held beliefs and probable sources of stress. Using good grammar, punctuation, and usage is both more important and less painful than most people realize!

Planning has a chapter all to itself. You may think you have established planning processes in your team or department; I'm sure most of you do. But how do you plan what you're going to write? Few people take a step back from the mass of information they could communicate to reflect on what they *should* communicate. This chapter will share tools and techniques to take control of your material before you take up a pen or touch a keyboard. Not only will it help you decide exactly the "big picture" questions of what to communicate to whom and for what purpose – it will make drafting the actual report much easier.

When it comes to writing, templates will not save you. We will look at the pros and cons of standard templates. How can you arrange a document to invite readers and increase accessibility? What software is available to help teams organize their working papers and produce reports? Should we be relying on software as much as we do – or more?

Finally...reports. This section is "Words into action." You may have expected the book to start with its ostensible subject. However, everything that precedes the act of writing is essential to its success. What is the purpose of reports? If they don't fulfill one, should we get rid of them? What do different professions and specialist areas – audit, risk, compliance, and IT – require or expect? Do international standards, where they exist, increase or diminish the pain of reporting? Some of the content may surprise you.

And truly finally...reviewing. It's a significant part of the pain of reporting, because it involves human beings, and their messy, complicated assumptions, impulses, and emotions. This takes us full circle: to culture, and the inherent difficulties in communicating with our own species.

The solution is not to leave everything to the available software. It is to show how putting this book's suggested approaches into place makes drafting and reviewing reports easier. Using principles and techniques from the world of publishing, you will see how to review your and others' work in a constructive, effective and efficient way.

If you want to change the way you write reports – for yourself and for your readers – then this book will help you. It will, in doing so, change how you think, how you communicate, and how you work.

And then what? Well, it will be a lifetime's work. No true writer is ever "finished," perfect or unable to learn more. The more you practice writing, and good writing, the more competent and confident you will become. This will not only improve your own morale, but that of your colleagues, as you share your expertise with them and thus improve everyone's performance.

You have nothing to lose except a few hours and many illusions. Enjoy!

NOTES

1 Thank you, George Orwell. "Politics and the English Language," *Why I Write* (London: Penguin Books, 2004), 105.
2 Oxford Dictionaries, https://premium.oxforddictionaries.com/definition/english/report and https://premium.oxforddictionaries.com/definition/english/radical
3 Steven Pressfield, *Nobody Wants to Read Your Sh*t* (New York: Black Irish Entertainment, LLC, 2016), 5.
4 The ratios are probably the same since the dawn of writing – a completely unscientific survey (my 30 years in academia, publishing, and industry) leads me to think that in any field, fewer than 10% of writers will be truly competent.
5 William Zinsser, *On Writing Well: The Classic Guide to Writing Nonfiction* (HarperResource Quill, 2001), 12.

Part 1

Words in the mind

Chapter 1

Communication and culture

Culture is more than important: it's essential to writing better reports. So, in addition to understanding the purpose of reports, we have to understand ourselves and our readers. What are our own beliefs and assumptions about reports, or indeed any workplace communication? Do we project these onto our readers? What are the results?

It's commonplace now to acknowledge culture as a factor, if not the major factor, in human interaction and indeed corporate life. The auditing of corporate culture has been topical and indeed fashionable in recent years, with teams looking at this complex subject in specific functions or departments. However, culture should be at the heart of most, if not all, investigations, audits, and reviews.

But what do we mean by culture in a corporate setting? People speak of an organizational or even team culture: "How we do things around here." And that's powerful – it will affect what people do, how they do it, and why. It's even more powerful for often being unspoken.

If we accept that every act of communication – and therefore every report – must unite feelings, thoughts, words, and actions, then we have to understand the culture or cultures in which that report is being written and read.

This takes us back to the purpose of reports. In theory, people write reports so that others can act on them, or gain new knowledge and insights. As we established in the Introduction,

Report:

- An account given of a particular matter, especially in the form of an official document, after thorough investigation or consideration by an appointed person or body.
- A spoken or written description of an event or situation, especially one intended for publication or broadcasting in the media.

DOI: 10.1201/9781003422365-2

However, to do so, we must worm our way into people's minds. They will be receptive to new information and exhortations to act only if they are willing to read what we write in the first place. They must want to read, attentively and through to the end.

Facts are not enough to prompt these reactions in people. As history – especially recent history – shows us, people can have all the facts relevant to a decision at their fingertips. They may choose to ignore them, pretend they don't know they exist, or even dismiss them as "fake news." What persuades people to acknowledge facts, to familiarize themselves with them, and, we hope, to accept them, is how we engage their feelings. How can we make them feel so that they *want* to read the report or other fact-based documents? How can we boost those feelings so that, even if the report contains unpalatable truths, they continue reading, and finally accept the facts as a basis for discussion and action?

If indeed we write reports so that people can act on them, we must realize that *facts are not the only factor*. Please note – I am not spouting the once-fashionable phrase "Feelings are facts." After all, if I am diagnosed with a severe illness but feel perfectly well, that doesn't mean I am not still ill. Feelings coexist with facts – they should not trump them, but often do when we don't acknowledge their existence.

Let's apply this to a commonplace topic that arises in many countries: that of how immigrant labor may affect a local population's chances of finding work. Imagine one group stating that immigrant labor doesn't affect the local population's chances, and using economic data alone to bolster its case. This group uses unvarnished facts, believing they are sufficient.

Imagine then an opposing group, claiming the discussion is not about economic data, but rather about their perceptions of the local economy. This opposing group may protest strongly that its members feel overlooked, marginalized – left behind. This group waves aside the economic data and says, "But this is how we feel!"

Chances are that the two groups will never hear each other, let alone come to any form of compromise. This is because neither is recognizing the other's position. In many cases, it may be that the empiricists – the fact-lovers – need to acknowledge the romantics' – the feeling-lovers' – position first. After all, it's hard to persuade people who feel no one listens to them if you don't listen to them.

This is incredibly difficult for empiricists to do – as one, I have often struggled with what seems to be someone's willful or egotistical insistence on feelings. "Your feelings are neither here nor there," I want to say. "Look at the video recording, or the phone-call transcript, or the automated test results!" I have never done this, of course, but have seen how even politely skimming over someone's feelings – "Yes, yes, very well, I think we've all heard your thoughts on that" – can shut down the conversation. "Well," the other person thinks, "if you don't want to listen to me, why should I listen to you?"

Again, it may feel frustrating for those of us who value facts and logic to feel we have to bow to the vague, touchy-feely world of the romantics. Yet without elevating emotion over logic, we *must* recognize that how people feel about our work affects how they read our reports.

The examples I have given are those of two sides in an economic and socio-political debate, and of two individuals in a work-based discussion. We can all, however, extrapolate to examples in organizations, where a department or team appears to fall into one or the other extreme category.

You may find – to create a completely stereotyped example – an IT team that lives and breathes binary code, and whose greatest passion is for the latest coding superstar in the field. To create another stereotype, you may work with a communications team that, while valuing performance statistics, focuses on how they and others feel about what they do. And both teams would be right to prioritize what they prioritize. In each case, this is their strength and purpose.

Understanding that is the first step to communicating effectively with them. If you understand their values, priorities, and assumptions, you understand their culture. You will therefore better understand what they do and why they do it – crucial when conducting any sort of review or investigation – and how to discuss it with them.

Given that you will have many discussions before writing any formal report, it is essential to get them right. If you approach verbal exchanges – whether in person, on the phone, or by videoconference – without considering culture, you may well end up speaking at cross-purposes. You may not ask the right questions, understand what's implicit in certain answers, or even get people to engage with you. The risks are clear.

Every act of communication – and therefore every report – occurs in *one or more* specific cultural spaces. Chapter 2 will explore team culture in greater detail, as well as profession- and specialism-specific culture. However, it's worth starting big, with the type of culture most people are born into, raised in, and often don't question.

NATIONAL OR REGIONAL CULTURE

This seems the most obvious one. Yet I've seen (and worked in) many organizations that pride themselves on being global, yet understand nothing outside their front door. (They often don't understand much within the organization, either, but we'll get to that point later.) You'll probably have seen business etiquette guides that acknowledge different cultural norms, which is helpful when attending meetings in foreign countries. So you'll have read of the importance of bowing in Japan, or how innocuous hand gestures in the US or UK could cause offense elsewhere.[1] It's great that those exist – at least you won't get run out of town five minutes into negotiations!

However, I'm writing this book a year into a pandemic that has brought business travel to an almost-complete halt. Whether national or international, meetings, conferences, and training that would have taken place in person now take place online. While this saves us jet lag, and helps the environment, it doesn't mean we can dispense with cultural considerations simply because we're not physically in the same room.

It never did.

Geert Hofstede's work on the role of national or regional cultures across corporations was new. It made many in business realize that, while they might not speak an overseas colleague's language, they could at least try to understand the colleague's likely values and assumptions. Hofstede's six dimensions model examines how different countries ranked in criteria such as individualism, power distance, masculinity, uncertainty avoidance, long-term orientation, and indulgence. The Geert Hofstede website[2] offers not only resources and tools for researchers, but also blogs and maps that anyone in an international organization can benefit from. His work has been valuable for people operating in different cultures, with his consultancy producing interactive tools allowing people to compare home and host culture. In this way, for example, a German person working in Dubai can see how the two countries compare in Hofstede's six cultural dimensions. This in turn should help the German national communicate more effectively and build good working relationships.

Given how acutely cultural differences inform communication styles, assumptions, and content, we must recognize them when interviewing, as well as when writing the resulting reports. One obvious difference is that between high-context and low-context communication. In low-context communication cultures, the listener or reader can take most words at face value. The context does not make a difference.

However, in a high-context communication culture, the listener or reader must look beyond the words for meaning. In a conversation, this means studying speakers for body language, facial expression, and tone to discern the "true" meaning of what they are saying. In reports, it means "reading between the lines" High-context cultures often present more opportunities for misunderstanding.

The following table illustrates perfectly what happens when a high-context culture (British) meets low-context culture (Dutch):

⚡⚡ ANGLO-DUTCH TRANSLATION GUIDE ▬		
What the British say...	**What the British mean...**	**What the Dutch understand...**
I hear what you say.	I disagree completely.	They accept my point.
With all due respect ...	I think you are wrong.	They are listening to me.
Oh, by the way ...	This is the primary purpose of this discussion.	This isn't very important.
I'll bear it in mind.	I won't do anything about it.	They will use it when appropriate.
Perhaps you could give this some more thought.	Don't do it, it's a bad idea.	It's a good idea. Keep developing it.
Very interesting.	I don't agree/like it.	They are impressed.
Could you consider some other options?	Your idea is not a good one.	They haven't decided yet.
That is an original point of view.	Your idea is stupid.	They like my idea.
I am sure it's my fault.	It is your fault.	It is their fault.

ELM EXPERTISE IN LABOUR MOBILITY   BY Nannette Ripmeester

Source: Nannette Ripmeester, Expertise in Labour Mobility

Created over ten years ago, this table (sadly, not always credited to its author) found immense popularity on the internet. People adapted it to their own experiences and professional contexts, with the most recent version I've seen offering "Anglo-EU" translation.

While the table is humorous, and not to be taken literally, it does speak to a deep truth. As a whole, British people – and yes, that term does denote a diverse group, including English, Scottish, Welsh, and Northern Irish – prefer a high-context communication style. It is common to find them making potentially unclear statements they expect other people to query, parse, or perhaps just intuitively *get*.

Given how many people must write reports in English, no matter what their native language, it's worth examining some assumptions. The anthropologist Kate Fox, in her book *Watching the English*, points to the cultural pitfalls inherent in many Britons', and particularly English people's, preferred communication style:

It takes foreign colleagues and clients a while to realize that when the English say "Oh really? How interesting!" they might well mean "I don't believe a word of it, you lying toad". Or they might not. They might mean "I'm bored and not really listening but trying to be polite". Or they might be genuinely surprised and truly interested. You'll never know. There is no way of telling: even the English themselves, who have a pretty good "sixth sense" for detecting irony, cannot always be sure.[3]

In contrast, a Dutch person (typically speaking perfect English) will leave you in no doubt that his or her "That's interesting" means the listener is indeed interested. That is because the Netherlands, for example, has a low-context communication culture. If a Dutch person tells you that he or she doesn't like your idea, your research results, or your criticism of his or her report...you'll know!

The Dutch are not the only people associated with a more forthright communication style. The journalist Emma Beddington wrote an amusing article entitled "What working from home has taught me about my partner and, worse, myself." In it, she observes her (French) husband's approach: "I note with alarm his tendency to speak his mind bluntly in professional settings instead of nourishing bitter, never-expressed grievances and sending overly polite emails, as is the British way. We must never be colleagues."[4]

Rinus de Hooge is a respected leader in internal audit, risk, change management, and counter-fraud. He has been working for different international banks, insurers, and asset managers for 30 years, with at one time 50 countries in his portfolio. A native of the Netherlands, he is skilled at communicating with international colleagues – often in English – and has first-hand experience of how mismatches in culture can impede communication. He is well attuned to the impressions people make on each other, and how their actions and assumptions can make or break relationships. His hobby – learning languages (German, English, Russian, Hebrew, and Arabic in addition to his native Dutch) – gives him further insight into cross-cultural communications.

Many native English-speakers do not adapt to non-native speakers' needs and assumptions – what's worse, they often won't even think of them, or of the gap that can result. As de Hooge says, this

> can create not only misunderstandings, but also create distance and make people become disconnected. My best connections, in English, are with non-native English speakers, e.g. my Eastern European colleagues. We are on the same level and there is no language "domination."

But he goes further to identify a common problem that will harm relationships even among people within the same culture. "The biggest barrier," according to de Hooge, "is that people are not prepared when dealing with each other – not knowing each other, not understanding each other's needs and not being interested in the human being."

According to the BBC, most native English-speakers – unless they have mastered one or more foreign languages – are the worst communicators in business.[5] The assumption that "everyone speaks English" and that everyone shares the same cultural references can leave overseas colleagues baffled and irritated. What's more, people who have never learned another language haven't had the experience of adapting their worldview, let alone speaking more clearly in a meeting.

It is of course impossible to overcome cultural differences, even if it were desirable. They often make for richer and more informed discussions. All we can do is be aware of our own assumptions and values, and that others' will differ. Emma Smith, Global Security Director at Vodafone, urges her team to do so by considering "the many different languages and cultures we communicate to. Slow down, use fewer colloquialisms and idioms, offer more time for questions." Even native English-speakers would appreciate this.

Since this book is in English, its focus is reporting in English. However, it includes advice and resources for non-native speakers writing in English, as well as for native English-speakers reviewing colleagues' work. It's common for native English-speakers to ascribe anything they disagree with in a non-native-English-speaking colleague's work to poor language skills.

Often, though, people who have learned a language to a high degree of proficiency are more correct in fact and expression than native speakers. This still doesn't stop some managers from relying on stereotypes and assumptions. As de Hooge points out, such complacency and disconnection go beyond potential misunderstandings in conversation. If we don't try to understand the culture in which our colleagues and clients operate, it's not only words and gestures we'll get wrong – it's entire documents and relationships.

It's important to understand others' expectations of what a report is and looks like, or even if it's necessary. Other countries may expect what may seem academic elements – references to scholarly studies, or footnotes, for example. They will differ in the degree of formality used in workplace communications. This means that using first names automatically, which is common in English-speaking countries, may not go down well elsewhere.

Another culture's formality may then be reflected in the register and tone of written communications such as reports – we may need to discuss how to produce a piece of professional writing that is suitably distant or formal, without being stilted or pompous. This is a hard balance to achieve. Finally, certain countries have a greater tolerance for lengthy texts, even while understanding that most readers would rather read a short one. If you are working with, for instance, a French colleague, don't be surprised if he or she sees lengthy reports as the norm. Such documents may be a sign of thoroughness and intellectual heft in that person's national and organizational culture. Again, simply being aware of the difference in expectations means you can anticipate and discuss any differences before they become misunderstandings or even barriers.

How we structure reports also reflects cultural expectations. We will cover this in greater detail in Chapter 6. Reports are usually most useful and reader-friendly if you think of them as newspaper articles; the headline should be clear, with the first paragraph giving the reader everything he or she needs to know. However, certain cultures – and organizations – prefer to start with positive messages, no matter what. We'll discuss in later chapters how this can mislead readers and potentially damage organizations.

Finally, people will differ in what they see as a logical order to the report; this may be down to national culture or simply a specific field. One cultural difference I have seen is in UK reports that may appear disorganized or repetitive to people from other cultures, especially North American and European ones. Having taught in universities, I was struck by the fact that UK (and Irish) students generally do not learn how to create outlines to plan their writing content. This is changing slowly but still seems limited to few people, often in post-graduate studies or on a specific studies-skills course. In most other western nations, children generally learn this skill in middle or early high school – between the ages of 12 and 14. This means they should have an advantage in quickly and confidently structuring whatever they write. It also explains what a German banker said to me when I explained this to him: "Ah! So this is why all the emails and reports from my English colleagues look like thought and word salads!"

It was a delightful image, but the message is serious. Anyone who doesn't rigorously plan their writing will confuse or at least irritate the reader. And cultural differences in education and professional training can exacerbate or mitigate the problem. Erin Meyer's *The Culture Map: Decoding How People Think, Lead, and Get Things Done across Cultures* is an invaluable resource for understanding cross-cultural communication in organizations. Particularly useful is her explanation of principles-first versus applications-first reasoning, which clarifies why people from different cultures fore-ground certain information, to the bewilderment of others.[6]

Even among native English-speakers, regional and national differences can affect how directly people speak. You could compare English and Scottish, Canadian and American, New Zealander and Australian; some would be more allusive in their communications, others more direct. As a native English-speaker, raised in the US but now living in the UK, I see certain public notices or email subjects as passive-aggressive or affectedly genteel: "Friendly reminder," "Please do not hesitate to contact me," or "Polite notice," where "Reminder," "Please contact me," or nothing at all would do the job. Having said that, even within the US, there are high-context/low-context regional styles – the classic example is the Deep South (extremely polite, to the point of being cryptic) versus New York City (direct to the point of brusqueness). It's "Well, I don't know that person, but I'm sure s/he's perfectly nice," versus "I hate that person."

I have also seen this play out in different ways in my own international career. Working as an academic and in publishing, I saw how Britons working in the US would mistake American directness as a license to be rude. Not having had, as Kate Fox would point out, the deep cultural grounding in being direct but polite, they would unwittingly offend colleagues, then wonder why. I have seen the same thing happen in banking – specifically, in internal audit, whose work needs to come across clearly in order to serve the organization. We'll see some concrete examples of this in Chapter 3.

SECTOR CULTURE

By sector, I mean private, public, or third sector. There are significant differences in how these sectors operate, and we should recognize them. They will certainly have different objectives and priorities, such as adherence to a government standard, making a profit in a new market, or providing help to people in need. However, it's not always as simple as drawing a line between the public and the private sectors – size, function, and organizational culture all play a role.

There is a common perception, especially among people who prefer less or smaller government, that the private sector is always more realistic and efficient than the public sector. I'm sure many of you have seen examples and counter-examples enough to know that this is simply not true – anything so sweeping rarely is.

However, perceptions are important, and professionals moving from one sector to the other bring with them preconceptions and stereotypes. They may feel that moving from consultancy to the public sector means they are bringing advanced thinking and bold innovation. The reality will depend on the size and culture of consultancy they worked for, and the public-sector organization they're joining. If the latter is relatively small, not obsessed with hierarchical formalities, and doesn't over-engineer its processes, the new employee may be pleasantly surprised by smoother or more efficient ways of working.

Nino Karazanashvili has held various public- and private-sector roles. Currently Head of the Internal Audit Department in MFO Crystal, a Georgian microfinance organization, she previously worked at the Georgian Ministry of Defense. Her roles there included Head of Standardization and Quality Control, and Head of Internal Audit. When I asked her what was the greatest difference in corporate culture, her answer was immediate: "In one sentence – independence in making decisions, which increases the responsibility." Comparing her public- and private-sector experiences, she says,

> In the previous context, in some cases (e.g. increasing human resources) there was no communication platform, while in the private sector I have this opportunity, but you need to be persuasive enough with your arguments in order to get desired results. In both cases you need to gain support, but in different ways.

Even within the private sector, professions and businesses are so varied, it is almost impossible to generalize. A global bank will have one culture; a car parts manufacturer another. The private sector can also include small start-ups run from kitchen tables, companies selling anything from cupcakes to coaching, cleaning services to consultancy. To speak of the private sector as monolithic overlooks the fact that these examples have vastly different strategic objectives, risks, and attitudes to risk. If we forget that, then we cannot provide useful insights and recommendations to our clients and colleagues.

The public sector too has a perceived image – sclerotic and bloated in some countries, trustworthy and citizen-focused in others. Yet it must depend on which kind of public-sector organization we're talking about: the National Health Service in the UK, the largest single employer in the country? Or a team of six people providing the weather service in a small nation? They will have little or nothing in common, except being taxpayer-funded.

John Chesshire is an experienced internal audit, risk management, and governance expert. He was Chief Assurance Officer for the States of Guernsey for over ten years, and has worked closely with government departments in the UK and overseas for decades. His view on the communication culture in government is that it varies hugely – there is no one-size-fits-all approach, simply because it is the public sector. "I think it's more about the size of the organization than the sector," he says.

> In my opinion, large, central government departments have more parallels with large, private-sector organizations than with small government agencies, or arm's-length bodies. Communications tend to be more formal, more hierarchical, bureaucratic and more rigid. Smaller organizations, irrespective of sector, tend to be slightly more innovative and more willing – and certainly more able – to adapt and evolve their communication styles, media and messaging.

So, if public-sector communication style depends more on size and function, and less on the sector, what are some constants? One may be the role played by elections. A challenge faced by national and local government departments is the possibility that, every few years, the people at the top will change. And with them, the strategic objectives and priorities. This means that people working in the public sector may need to completely change direction, possibly abandoning projects they have worked on for years. In turn, this means that audit, risk, or compliance reports will have to reflect the change in strategy, objectives, and priorities – all while making their work available to the public.

Many public-sector bodies and organizations have been urged to imitate the private sector over the past few decades. Even the financial crisis of 2008 seems to have done little to dent the myth of superior management and operations in the private sector. Universities in the UK, which traditionally rely primarily on public funding, have had a crash course in private-sector attitudes since the turn of the century. After the rising number of higher-education institutions and the introduction of tuition fees, UK universities exist in an environment unimaginable a generation ago. The world of business has entered the world of academia, with mixed results. The notion of a competitive market sees universities vying for student applications. The insistence on quantification and especially key performance indicators

(KPIs) has led, many would say, to a focus on what can be counted, rather than what counts. All of this has required increasing managerial, rather than academic, positions so that the real growth in universities is bureaucracy rather than research and teaching.[7]

Unsurprisingly, this affects communications. Having worked with both academics and people in industry, I was used to the different cultures and styles. However, when working with university administrators, what I saw was the result of fundamental misunderstandings about written communication. As André Spicer has pointed out, "What is striking is not just that universities have taken so keenly to the language of business, but that they have been so routinely bad at it."[8]

Many people refer to "academic writing" and "business writing." They often imply that the former is obscure and pretentious; they don't realize how often the latter is, too. Many university administrators, neither academics nor people with any experience of business, often feel compelled to prove their ability to master both "academic" and "business" writing. I will let you imagine the result. The point is that no one should obsess about aping "academic" or "business" writing. The focus should be to produce *good* writing.

Finally, the third sector – often the charity sector – is caught between public and private in many ways. It relies often on government grants and private donations, but also increases revenue through sales. Pet charities sell calendars; stately homes sell tickets to visitors, and earn money through cafés and gift shops. In this sense, they are akin to many universities, in that they must somehow combine public- and private-sector objectives and practices. However, the underlying ethos may complicate such a pragmatic approach.

Jane Bettany is Head of Internal Audit for Frontline AIDS, an international charity. According to her, the sector attracts "highly engaged staff who really care about the organization and the cause. The pay is lower so they tend to be less money-driven and more cause-driven." This focus on the cause rather than material rewards means that people feel personally engaged in their daily work. As a consequence, what may be routine tasks in other sectors or organizations speak here to people's hearts and identities. We'll talk in Chapter 2 about how this can affect communication within and among teams, when "personal" and "professional" combine.

Since Frontline AIDS is an international team, Bettany is mindful of broader cultural differences, as well as sectoral. She considers not only the fact that colleagues are often personally committed to the cause, but also the fact that they have different communication styles.

> I am aware of different needs, to be formal or more polite sometimes, or more friendly, depending on the country. We work a lot with Africa where people are so very warm, calling each other friends or colleagues, and I forget to add the warmth to my emails! Even legal letters sometimes have much more emotive and descriptive language in them.

Mark Carawan is Senior Fellow at New York University School of Law, Program for Corporate Compliance and Enforcement. His previous roles include Chief Internal Auditor at Barclays and Citi, and Chief Compliance Officer at Citi. He echoes Bettany's sentiment, and underlines how important it is for organizations to develop a communication strategy that works for everyone. "The communication style must be respectful, sensitive and adaptive to get the message across. A communication style which does not adapt is ineffective, probably offensive, and at odds with an ethical culture, whatever the jurisdiction."

And so we return to national culture. Culture never walks alone – we are the sum of our influences, from national through sector-specific, to professional and organizational.

PROFESSION- AND SPECIALISM-SPECIFIC CULTURE

Depending on your education, professional credentials, and further development, you will already bring many habits, standards, and assumptions into your role. We often think of traditional professions such as law and medicine, and their accompanying jargon: legal "terms of art," for example, or the Latinisms of medical terminology. However, in audit, risk, compliance, and IT security, a range of specialist training and conditioning can feature in our reports.

In audit, for example, different backgrounds and qualifications dominate in different countries. External auditors are still usually chartered accountants, and, given external audit's focus on accounts, that makes sense. Government auditors often have similar qualifications, too, and this links directly to value for money: taxpayers want to know what their national and local governments are spending taxes on. In fact, until recently, it was common for internal auditors in many countries to come to the profession from an accountancy background. This is clearly useful when assessing financial controls, especially in countries with stringent financial regulations. It's obvious why any organization will want to include accountancy specialists in the team.

However, accountancy is only part of the picture. Qualifications specific to internal audit equip internal auditors with the knowledge and skills to assess a range of internal controls – not only financial ones. The emphasis on risk-based internal auditing, which focuses on those things that could prevent an organization from achieving its strategic objectives, clearly goes beyond traditional "bean-counting." A strong internal audit team is one that understands the broader organization and the range of relevant risks and controls. This means a diverse internal audit team will see more, understand more, and communicate more effectively.

So is there a single internal audit-specific culture? Traditionally, internal auditors were seen as interested only in numbers and "box-ticking" – jokes about internal auditors coming around with their clipboards were the norm. There was some truth to this, especially in internal audit teams that conducted cyclical audits. They would audit payroll every year, procurement every 18 months, and recruitment every two years – or something like that. If a new payroll system drastically reduced the risk of manual errors, it didn't matter: the team would continue to audit payroll every year, no matter what. Meanwhile, over in procurement, a new senior manager dispensed with the existing process for assessing bids, in favor of a "streamlined" approach that allowed him to approach, assess, and approve suppliers with no checks or balances. Would it still be wise to wait before auditing that area?

Risk-based internal auditing brings the emphasis squarely back to risk. What is it, where is it, how do we assess it, and how can we mitigate it? This requires not only understanding risk, but also strong communication skills. Internal auditors need to work closely with other assurance providers both inside and outside the organization. Internal teams include risk, compliance, regulatory reporting, and subject-matter experts such as cybersecurity analysts or project managers. Outside the organization sit external auditors, consultants, professional bodies, and regulators. How these different specialists communicate with each other can make the organization stronger – or more vulnerable.

Having looked at organizations inside and out, Mark Carawan sees internal and external auditors as having more in common than many may think.

> I do not think there are any inherent cultural differences between external auditors, internal auditors and compliance officers: each role demands adherence to principles requiring performance of fact-based assessments, transparency, integrity, and independence. Their roles are different, but there is a professional culture which binds them together.

This point takes us back to that made earlier: we need to understand the culture in which we work, interrogate our own assumptions, definitions, and priorities to make them explicit. Do all those who share a professional culture speak the same language? They may, but we will find out only by asking. Then we can say, for example, "When I talk about risk, this is what I mean." And even more importantly, "What do you mean when you talk about risk? How do you see it?"

If we are speaking with a close colleague, we are likely in agreement, and can exchange reports using that term confident that we share the same understanding. However, what happens when we send that report to someone in senior management? First, let's go back to basics: who are we sending it to? What is that person's background? How could he or she interpret what we have written? If the person comes from project management, the word "risk" may conjure up a far broader meaning: anything that could delay the project. If we send it to someone in cybersecurity, that person may think we have overlooked many worrying possibilities labeled "risks" in their team.

Internal audit vocabulary comes to seem intuitive to practitioners. To anyone outside, or to novices, it can seem oddly clinical and even grudging. Consider the terms "adequate" and "effective." To internal auditors, the terms are clear. An "adequate" control is one that is designed in such a way that, properly performed, it will mitigate a risk to acceptable levels. A classic example – one often used when training internal auditors – is that of crossing a road. Adequate controls would include visible crosswalks, functioning traffic lights, accepted rules of the road, and pedestrians' understanding they must look both ways. (Bear with me – this example is useful and will come up in later chapters.)

By "effective," internal auditors mean that people follow the process or adhere to the control. In our road-crossing analogy, the controls are effective only if people use crosswalks, cross on the light, follow the rules of the road, and look both ways. If they don't do any of these, then the controls are adequate but not effective.

These are important concepts in internal audit communications. If we say in a report that a control is adequate but ineffective, it means that it's a good control – but no one is following its directions. If we say a control is inadequate – well, there's no point seeing if anyone adheres to it, because even if they did, it wouldn't mitigate risk.

However, can we rely on readers to understand what we mean by the terms "adequate" and "effective"? Given how important the concepts are, we must take care to explain our terms to a wider audience. Suppose a report states that a control is "adequate" without defining the term. Senior managers reading the report may take umbrage, believing the term damns their work with faint praise: "Merely 'adequate'? That's not very encouraging – we worked hard to develop a sound process." After all, imagine that your manager told you your work was "adequate" – what would you think? This isn't about being wordier, or adding glossaries to reports. It's about being conscious of gaps between our own understanding and other people's, and working to bridge them.

It is crucial to define and articulate any concept we see as fundamental. This is not easy – it requires alertness to ideas and tasks that have become second nature in our profession or specialist area. However, the very fact that it's second nature – that we use or do something without thinking twice – means it may be unclear to others. "It goes without saying" or "It's self-explanatory" are risky in any field – all the more so when discussing topics that could help or harm an entire organization.

Most people want to see their organizations not just survive, but thrive. For those of us tasked with keeping organizations safe – through monitoring, assessing, and reporting on risks and controls – clear communication, first with ourselves, then with others, is crucial.

ORGANIZATIONAL CULTURE

This chapter has so far covered an array of cultures, all of which affect communication. National and regional differences are what most often come to mind when people speak of cultural differences, and it's clearly important to try to bridge gaps.

Further differences arise within sectors. People working for government will have different priorities from those working in a global bank, or a charity. However, the lines between sectors are not impossible to cross. Similarities in size and function can create common attributes, so that, for example, a two-person risk team in a local council will be able to share similar experiences with a three-person compliance team in a small credit union. The two teams may be in different sectors, with different specialist backgrounds and qualifications, but they can still find common ground.

Organizational culture is what dictates many people's daily lives. Whether large or small, organizations tend to have "personalities" and attributes, which are nothing to do with public image. The organizational culture is what people mean when they talk about "tone at the top." Culture usually flows downward, from board level, and it is the executive that decides the organization's values, beliefs, and priorities. As with national culture, organizational culture likely affects every aspect of people's experience in that entity.

In organizations, a good or bad culture dictates what people do at every level; this includes communication. Chapter 3 will give specific examples of how communication both promotes and reveals the true nature of a company's culture, through real-life case studies.

How and what an organization communicates displays its culture. What does it believe is important? Is it willing to say so publicly? What constitutes success in the organization, and what does senior management allow or encourage to achieve it? Is failure an option? Since failure is inevitable in any human endeavor, how does the organization talk about it? The answers to all these questions will tell you much about the culture.

Often, national and organizational cultures reinforce each other to hide problems. It's always worth asking why. If a problem has arisen, but nobody is willing to say so, does the problem still exist? Of course it does. This isn't a philosophical conundrum about trees falling unheard in the forest, or a single hand clapping. This is about whether professionals communicate as adults, or as children, covering their eyes in the belief no one can see them.

Let me share one example from my experience as a report-writing trainer and consultant. A financial services organization's risk function asked me to review reports and deliver training to improve them.

One report in particular puzzled me. I read from the beginning and found the executive summary reassuring. The team had reviewed a high-risk area and seemed to have covered all the relevant controls. The report spoke of how hard senior managers in this area had worked, building relationships across the business, and improving processes by streamlining them.

Several pages on, the individual findings began. There were quite a few, couched in cautious, diplomatic language. By the third finding, I began to feel uneasy. There were hints – nothing explicit – that senior managers had perhaps streamlined processes by circumventing controls. This was not what I had expected from the executive summary, which had in no way suggested that the improvements included diluting or removing basic checks and balances.

Reading further, I saw more and more findings that not only suggested deliberately weakened controls, but also potential fraud. The same senior managers who had streamlined controls seemed to personally benefit from the changes, whether through bonuses or even lucrative conflicts of interest.

This was a classic case of "reading between the lines." There was nothing explicit or overt, simply page after page of heavy hints. I asked the team what was behind this. Was I cynical after years in internal audit? Or was there indeed misconduct? The exchange that followed was extremely revealing, and went something like this:

ME: Um, it reads as though the senior managers are gaming the system here.

TEAM MEMBER 1: Um, yes. That's what we thought.

ME: OK… So, why so coy about it in this report? Or did you already cover this in reports to the board and risk committee? And I'm assuming you notified the police.

[silence]

ME: Anyone? What happened?

TEAM MEMBER 2: Well…it's tricky.

ME: Meaning?

TEAM MEMBER 1: Well…it's delicate. It's a difficult conversation to have, accusing people of fraud.

ME: OK, I understand. Are you saying that you didn't have any hard evidence?

TEAM MEMBER 3: Oh, no – we had loads! A stack of paper about a foot high. All corroborated.

ME: Given all this evidence, you must have notified the board, risk committee and police, right? You surely didn't leave it until this report, which is just a routine review, did you? I mean, this report doesn't even refer directly to what you've just said happened.

[silence]

TEAM MEMBER 2: As we said, it's tricky. I mean, you don't want to have awkward conversations and damage relationships in the business, do you?

ME: Sorry, I don't understand. Which relationships would be damaged by reporting this crime?

TEAM MEMBER 1: Well, the senior managers in the area, of course.

ME: In other words…the fraudsters?

[silence]

The head of the risk function was at this discussion, looking troubled. I spoke to him afterward, and what he described was a perfect storm of cultural conflict. He had spent seven years in the US before returning to lead this team in his home country, and had brought back with him more direct ways of expressing himself.

He observed, he said, a clear cultural difference between his US experience and his current one – one he never would have questioned before working abroad. However, his team's unwillingness to state the truth openly was not simply down to cultural differences in communication style; it was a direct result of the organizational culture.

In organizations that "shoot the messenger," problems are swept under the carpet. People working in them quickly learn to either conform or leave. (This head of risk eventually left.) The board and senior managers may feel that whatever they are doing is worth the risk. They may eventually have to pay a regulatory fine or settle a lawsuit if caught, but may feel that any penalties are outweighed by profits. The broader ethical question doesn't seem to come into the equation.

Other organizations may unwittingly create the perfect conditions for cutting corners and even committing fraud, even if those at the top don't intend it. Companies that pride themselves on speed at all costs, innovation no matter what, and short-term results can put pressure on staff to do whatever it takes to meet targets.

In any organization, the language people use, and how they communicate, reveal the culture. Do they use euphemisms and copy in their entire hierarchy to any email? Does every report take a month to finish because of the layers of review and senior managers' need to rewrite everything? Do internal communications encourage buzzwords and trendy phrases at the expense of clarity and precision? Is everyone under pressure to produce by sometimes arbitrary deadlines, with little time for reflection or debate? Is there any opportunity to persuade or influence those higher up?

If any of this sounds familiar, it reveals the organizational culture you work in. You may even think that since the company is doing well, it doesn't matter. The profits are steady, so this way of doing things must be sustainable. Well, possibly – but only for so long, as we will see in Chapter 4. There is a difference between doing good and doing well; the words organizations use reveal what matters most to them.

ACTIVITIES

- Consider your current role and organization. How many cultural influences can you associate with them? For example, if you are a Canadian IT security manager working in an Irish bank, you may have the following elements:
  - Canadian upbringing and education (English, French, or bilingual?)
  - Education and other experiences outside Canada
  - Professional training, standards, and expectations as an IT professional
  - Irish cultural expectations (again, dependent on location within Ireland)
  - Bank's corporate culture
  - Bank's security risk appetite or tolerance
  - Team-specific culture (Is the focus on coding skills to outwit computer viruses? Is it on customer communications? Is it on supporting the group risk function in its reporting?)
  - Anything else?
- Now consider which of these elements is a strength or weakness in your communications? Do any of them contradict each other? For instance:
  - You may have been raised to be direct in a French-Canadian environment; could your Irish colleagues find you too blunt? Or could they learn from your clear communications?
  - Your IT training is likely to have focused on quantifiable results. Is this useful for reporting to senior decision-makers, or does it dilute the "big-picture" messages?
  - If you have worked in other countries or organizations, does this help you see the bank's corporate culture more clearly? What works well in it? What could be better?

You can approach these activities however you wish – you can simply reflect for a few moments, or document your thoughts. However much or little you do, the simple fact of thinking about culture and how it affects communication will help you draft your reports with greater awareness of your readers and yourself.

SUMMARY

- Different values and approaches – from national to individual – inform both verbal and written communication. Culture isn't only about language and gestures. It's about how people perceive the world and their place in it. It affects how they communicate, and how they receive others' communications.
- People working in public-, private-, and third-sector organizations may all have different approaches. However, the size and function of an organization can blur the sector lines: small government agencies may feel alienated from large government departments, while having much in common with a small- or medium-sized enterprise with a comparable number of employees.
- Professions have their own standards, qualifications, and language. This in turn informs not only how they communicate, but also the prism through which they see their professional world. It is not surprising if an auditor and project manager have different definitions of the word "risk," or if accountants and cybersecurity analysts quantify things differently.
- Organizations may not be people, but they may still have "personalities." When people talk about "tone at the top," they talk about the values and beliefs – the culture – within the organization.

NOTES

1 For anyone who has heard the myth that communication is mostly non-verbal, here's a useful corrective: David Lapakko, "Communication is 93% Nonverbal: An Urban Legend Proliferates," *Communication and Theater Association of Minnesota Journal* 34 (2007): 7–19.

2 https://geerthofstede.com/culture-geert-hofstede-gert-jan-hofstede/6d-model-of-national-culture/

3 Kate Fox, *Watching the English: The Hidden Rules of English Behaviour* (London: Hodder & Stoughton, 2004), 181–82. See also David Shariatmadari, *Don't Believe a Word: From Myths to Misunderstandings – How Language Really Works* (London: Weidenfeld & Nicolson, 2019), 209.

4 *The Guardian*, November 30, 2020, www.theguardian.com/commentisfree/2020/nov/30/what-working-from-home-has-taught-me-about-my-partner-and-worse-myself

5 Lennox Morrison, "Native English speakers are the world's worst communicators," *BBC Worklife*, October 31, 2016, www.bbc.com/worklife/article/20161028-native-english-speakers-are-the-worlds-worst-communicators. Furthermore, in an article in the *Financial Times*, Gillian Tett suggests that British workers could increase productivity by communicating more directly. "It's time for us all to speak more like the Dutch," February 29, 2022, www.ft.com/content/29d360f8-fa26-4ade-adb8-298cb2ee1913

6 Erin Meyer, *The Culture Map: Decoding How People Think, Lead, and Get Things Done across Cultures* (New York: Public Affairs, 2015). 89–104.

7 "Today, two thirds of universities employ more administrators than they do researchers." Hettie O'Brien, "From bogus self-employment to zero-hours contracts – why working life has become unfulfilling," *The Guardian*, June 19, 2021, www.theguardian.com/books/2021/jun/19/lost-in-work-by-amelia-horgan-review-why-so-many-people-feel-unfulfilled. See also "The Guardian view on funding universities: the market model isn't working," *The Guardian*, May 30, 2021, www.theguardian.com/commentisfree/2021/may/30/the-guardian-view-on-funding-universities-the-market-model-isnt-working

8 André Spicer, *Business Bullshit* (London: Routledge, 2018), 95.

Chapter 2

Communication within and among teams

Understanding, not understanding, or simply ignoring someone's national, regional, professional, or organizational culture can be the difference between success and failure.

This doesn't mean, though, that once people work in the same organization, they share the same assumptions, values, and objectives. They'll probably have many in common: an understanding of what the organization does, and a desire to see it succeed, for example. But once people sift into different divisions and teams, differences become apparent.

These differences can be down to different locations – a global organization will have locations in various countries, which brings us back to culture. Even an organization limited to one country will see different approaches and styles in different locations; staff working in New York City will not have the same culture as those in Minneapolis, and Londoners may find they have little in common with Glaswegian colleagues.

The distinctions can be profession-specific, too; group legal and group HR are usually different creatures. However, even within those two groups, communication doesn't always flow.

DOI: 10.1201/9781003422365-3

Source: Virpi Oinonen (businessillustrator.com).

WITHIN

Many colleagues will likely share the same educational or vocational qualifications, professional standards, definitions, assumptions, and objectives.
Despite this, different roles and personalities mean it can often be harder to
get immediate colleagues to sign off on something than to get "outsiders"
to agree to it.

What kinds of communications take place within teams, how, and why? If
we examine these points first, it should then shed light on how miscommuni-
cation arises among like-minded people, even before other teams join the fun.

In an office, face-to-face communications are usually informal conversations. There will be regular team meetings, but most interactions will be
across the desk, passing in the hallway, or in a shared kitchen area. They are
relatively unstructured, create either good or bad personal relationships at
work, and so inform how people will interpret their interlocutors' written
communications – including reports.

Email, the bane of most people's lives, lends itself to virtually any communication form or style. It can be a brief personal query about a colleague's weekend, or a lengthy, structured, more formal document – almost a minireport, but one using the medium of email. Again, the ease with which most people can send emails means that they are less likely to think about what they are saying, why, or to whom.

Why does this matter? First, bombarding colleagues with pointless emails is irritating. It is the equivalent of perching on a busy colleague's desk and chattering away about something trivial. Most emails aren't necessary, and copying multiple names into emails is a sign of defensiveness. The senders are either deeply insecure, or need to protect themselves in a rotten corporate culture.

Second, emails – like casual conversations – set the tone for our written work. This can be good, as when we ask a trusted colleague to look at a rough draft. This colleague, knowing the writer and the context, is likelier to read the draft with a sense of perspective. He or she will be less likely to criticize every minor error and will focus on what is most important to helping the writer progress. When we know and trust our colleagues, we tend to be better at giving and receiving comments about our work.

However, this familiarity can also mean that we forget to include important information, or that we become sloppy and burden colleagues with unnecessarily clumsy work. John Chesshire refers to clumsy writers who subject him to such drafts as "serial writers" – but not in the sense that they're writing a hit television series. These are the people, he says, who keep coming back with ever-more verbiage to inflict on the poor reader – the victim. "They're ratcheting up the torture and the body count. Often I feel as though I've been bludgeoned, because I cannot merely skim – I have to read every draft all the way through. Afterwards, I feel as though I have a brain injury."

This isn't a flippant comment; often, when reading reports, I too have experienced worrying symptoms of an impending migraine or worse. Visual disturbances, looming pain, especially on one side of my head, nausea – all of these have happened simply from reading and re-reading extremely badly written reports. Too many clients, especially senior managers tasked with reviewing draft reports, have told me the same thing. It's not actually surprising – between the eye strain of repeatedly staring at a page or screen, and the mental strain of trying to extract some meaning from the impenetrable, our bodies tell us it's too much.

A senior governance professional recently asked me if I thought he had dyslexia. I'd seen no evidence of this in our work together, so asked him why this had occurred to him. His answer was that when he reads reports at work, the words blur and jumble on the page. Because of this, he assumed he had a problem. I asked him if he experienced this when reading anything else – such as a newspaper article, novel, or even instructions for a toaster. The answer was no. This demonstrated perfectly how bad writing can affect readers mentally and physically.

These physical reactions occur when we are trying to be attentive and rigorous in our reading. However, time, energy levels, and shared assumptions often conspire to create another situation. It's less physically painful but counter-productive. This is when we assume near-complete understanding from our closest colleagues. Imagine I see someone every day at our shared workspace, that we chat together at the coffee-maker, and that we give each other brief, informal updates throughout the day across the desk. We will naturally assume that each has all the necessary information and context.

We may have much of it, but we're not inside each other's heads. So that means it's all too easy to knock out a draft that makes perfect sense to us, pass it to our "work wife" (or husband or partner – our other half in the office), and expect total understanding. You have probably experienced this yourself, and at least one of the two things is likely to ensue. Maybe the colleague won't understand why you leap from A to D, or which standards are underpinning your risk assessment. This will take further time to explain, which means your review process isn't as efficient as it could be.

Familiarity with one's work and colleagues means that seeing it out of context can come as a shock. When I ask people to rewrite work they're often extremely close to, utter bafflement is common. When they actually recall what they meant to convey, their rewrite bears little resemblance to the original. This brings home how dangerous it is to remain inside our own heads, whether individual or collective.

The second scenario commonly occurs in tight-knit teams, where the colleague you ask to review may truly be close to your inner thoughts. If this is the case, he or she will make the same assumptions and overlook the same gaps. Now we've traveled from extreme irritation, eyestrain, and headaches, through inefficiency because of further explanations, to groupthink. Some would say that groupthink is the riskiest of all situations within a team.

Margaret Heffernan's book *Wilful Blindness: Why We Ignore the Obvious* explores all the ways in which human beings unwittingly and often deliberately avoid seeing what is in front of them. She examines several cases in which companies failed because of groupthink, where only one approach, vocabulary, or version of the truth was acceptable. In many cases, individuals knew things were going badly, but didn't speak out. They didn't know how to without alienating their group, their corporate "family." They feared losing promotion or even their livelihoods. In many cases, though, they had simply fallen out of the habit of questioning or contradicting colleagues – which, as Chapter 1 showed, is even harder in some cultures than in others.

Heffernan addresses the common human experience of living and working among those we resemble. Yes, it is natural to seek out our fellows, and it can be comforting, but "One of the many downsides of living in communities where we are always surrounded by people like ourselves is that we experience very little conflict. That means we don't develop the tools we need to manage it and we lack confidence in our ability to do so."[1]

Given how many waking hours people now spend in the workplace, compared to at home or with friends and neighbors, it's not unreasonable to see our team at work as our community.

Any of these situations, however, can be resolved through goodwill, as well as heavy doses of time, patience, and energy. In the real world, though, few of us have such gifts. As a result, things can often go awry when the very familiarity we think will create the perfect conditions for effective communication backfires.

Many people have mentioned to me that their closest colleagues can be the greatest sources of friction. We may often think that our team and function will present a united front to "outsiders," whether clients, senior managers, or board members. We want to make sure we all share the same understanding, and agree on how to articulate it, before sending a report. However, this isn't always the case.

There can indeed often be more discord and tension within the team producing the report, than between that team and the report recipients. What contributes to this, and how can we address it so that we spend our time producing better work for the organization, rather than squabbling with our closest colleagues?

Let's first recall the earlier points about culture. Even if everyone in a team shares the same nationality, regional differences may mean conflicts in communication. Furthermore, differences in background, education, professional orientation, and further specialization mean colleagues working in the same team can be far removed in their assumptions and priorities.

Within audit and risk teams, for instance, you are likely to find people who have qualifications in those specific fields. However, many come from completely different backgrounds: accountancy, finance, law, IT, procurement, even former academics in the sciences and humanities. While this means that large teams with diverse members will be able to draw on a variety of knowledge and experience, it also means they must be conscious of the need to foster communication within the team. Otherwise, the risk is too high that different people's unspoken assumptions will surface exactly at the wrong moments. Different understandings of crucial concepts such as organizational strategy, risk appetite, or materiality thresholds need to be articulated and reconciled before the moment when they could derail important work: a crucial client meeting; the organization's risk register; or a report to the audit and risk committee.

Rinus de Hooge gives a specific example of having to clarify a term everyone in a meeting assumed they had understood.

I performed a strategy audit in a Dutch environment and started asking what people thought of strategy. It turned out that, after some digging into the replies, everyone had a different definition of the word strategy. For some it was about "getting things done," for others it was "about a goal," and again others had "a road map" in their head. Others used the word "strategy" with activities and mingled them. Unclear, unspoken, different combinations of definitions, implicit in people's heads, made connecting difficult. It had led to non-implementation of strategy, as no one was quite aware what it meant.

It's clear from this example that even within the same national culture, within the same organization, people have different interpretations of what should be a clear, commonly defined term. The result in this case was that the organization hadn't actually put its intended strategy in place, doubtless thinking that the term "strategy" spoke for itself.

Another example is one I saw repeatedly in the internal audit function of a global bank. Most of the people working there were well qualified, knowledgeable, and experienced. However, they were also human. It was therefore common for them to assume that their own understanding was universal, which led to some risky situations. One I am thinking of in particular was when teams from low-context cultures – the US, Netherlands, Germany, and German-speakers in Switzerland – would submit reports to their functional heads in London. The London-based heads would then change the style of the reports to suit their own perceptions of what was "correct." This led to two problems.

First, the UK teams would assume that their overseas counterparts, "not being English," needed to have their reports rewritten. However, given that the overseas teams overwhelmingly had much better grammar, spelling, and punctuation than their UK colleagues, this step simply introduced basic errors into what had been well-written reports.

More serious still were stylistic changes – those palatable to a high-context culture such as the UK's. Where reports had been clear and direct, while still professional and respectful, they were now vague, wordy, and evasive about what, if any, problems existed. As a result, the reports lost their value to the readers. Not only were they harder to read, they also obscured necessary information.

There were further consequences to these changes, borne out of cultural differences. One was that UK teams did not appreciate the extent to which other countries' regulators would disapprove of reports that appeared evasive. Some countries' regulators were known to be much tougher than the UK's. They were readier to issue heavy fines and start criminal proceedings, and in some jurisdictions remove auditors' license to practice.

People in the London office had assumed that because they and their colleagues around the world worked in the same specialist area of audit, they worked to the same rules. This meant they weren't aware of significant differences in the team: cultural, certainly, but also regulatory and legal. This could, if allowed to continue, have landed some non-UK colleagues in hot water in their own jurisdictions, with severe professional, financial, and possibly legal consequences.

Within teams who share a common background and understanding, there is still potential for discord. This can arise from different perspectives and deeply held views. As Jane Bettany says of her observations of the charity sector,

> Everyone has an opinion that is sometimes part of the core of who they are, so dissenting decisions can be difficult to resolve. Decision-making in some organisations can tend to be more collaborative or consensus-driven. This takes a lot more time and can in the end upset a lot more people because they were given a chance to voice their opinion, but then don't feel they were listened to if their view wasn't taken forward.

What Bettany describes so eloquently is not limited to the charity sector. Many people in public- and private-sector organizations are deeply attached not only to their own views, but also to their particular specialist angle. Thus someone from a risk management background may feel unheard if a colleague advocating entry into a riskier, but more lucrative market, carries the debate. This in itself may arise from misunderstanding the nature of debate and discussion. Disagreeing with someone is not necessarily personal, and understanding someone's position doesn't mean agreeing with it. National or regional cultural differences can make these nuances even harder to grasp.

Many believe that the size of the team or broader organization can play a role, and John Chesshire described this dynamic in Chapter 1. However, humans are complicated – sometimes people bring a communication culture with them, wherever they go. Ola Bello is Head of Internal Audit at a medium-sized, UK-based insurer with approximately 500 employees. She previously worked in the insurance internal audit division of a global bank with tens of thousands of staff, but didn't change her approach when she changed organizations. "My communication style doesn't seem different with both firms. I think it depends on individuals rather than a firm style. Things like familiarity with the contact or stakeholder." For Ola, clearly, the individual in front of her is what determines the exchange – not the size of the company.

The 2020 pandemic saw people across the globe move from office-based work to remote working. Whether in large or small organizations, video-conferencing became the norm for many meetings, training events, and even regular one-to-one updates. Most people believe that remote working will continue in some form, even if only part-time, as it has proved generally reliable and cost-effective. However, the effects on communication within teams have varied widely.

Many people are keen to return to offices, citing "screen fatigue." (Some even admit to missing their colleagues!) Others have been pleasantly surprised by how remote working has broken down barriers. Previously, it may have been difficult to get even a 15-minute in-person appointment with a senior colleague. When people work at different locations, travel to conferences, or are away on holiday, it can be difficult to coordinate diaries. With reductions in travel across the globe, many more people are at home and more available to accept a videoconferencing invitation. In turn, this has increased the amount of verbal communication at all levels.

A series of webinars run by the Chartered Institute of Internal Auditors in 2020–21 showed that many members had benefited from being able to speak directly (via telephone or videoconference) with audit committee members and other senior staff in a way that previously would have been difficult. This in turn affected their reporting, often for the better. Having developed better relationships with people across the organization, the internal auditors felt less need to "over-engineer" their written communications. Reports in many organizations became shorter and more to the point, thanks to regular verbal communication establishing common ground. Previous reports – often lengthy and full of irrelevant detail – had arisen in a context of little face-to-face (including virtual) interaction and thus a wide margin of error in communicating.

Many senior managers realize the potential for misunderstanding in their own teams. After all, they are the ones who must first review and decrypt draft reports, well before other teams or clients see the final version. I firmly believe that 90% of my clients have come to me out of desperation – a classic "burning platform" situation. They don't want to spend budget on training or consultancy, but it seems less painful than continuing as they are. The former may just work; the latter is clearly not an option anymore.

However, an alarming number of them overlook more serious risks arising from their existing reporting process. They may see report-writing training as a way to make the team work more efficiently, and review less painful. Many will see it as an opportunity to communicate more clearly with colleagues and clients. Vanishingly few realize the financial, regulatory, legal, and reputational peril their existing reports already put them in, simply through poor communication.

How then do those intra-team communications affect inter-team communications?

BETWEEN OR AMONG

We've seen the scope for misunderstanding within teams that supposedly have a great deal in common: professional qualifications, objectives, working processes. However, when team members share a common understanding of terms and reporting, it can lead to groupthink. Not only can this stop people from seeing a bigger picture, with different elements and risks. It also increases the chance they will communicate at cross-purposes with "outsiders," anyone who is not part of the team.

This cross-communication can take several forms. Most obviously, it can alienate or confuse people. However, as stated above, it can pose an even greater risk: if it is not clear what different teams mean by something (such as "strategy," "risk," "material"), readers outside the team may believe they understand perfectly. But they won't.

One group in particular seems to be extremely aware of its specialist terminology and therefore the risk of misunderstanding. When working with IT specialists, whether auditors, cybersecurity analysts, or business continuity consultants, I'm struck by how they describe themselves. They know they use terminology specific to their field and that non-specialists can feel intimidated. They also know they have to take pains to be clear and simple in their explanations. However, I feel they often focus on the wrong things.

It's ideal to be aware of the gap between one's professional group culture and others. It's another to feel you must constantly apologize for it, or agonize over terminology that may be precise and useful. All too often, IT specialists worry about wording that either has become commonplace or is self-explanatory. Local or wide area networks; parallel testing; log-in credentials; phishing – most senior managers will immediately understand these terms. Instead of focusing on what isn't a problem, why not focus on what is?

One problem IT specialists and indeed entire organizations face is project management-speak, acronyms, and business buzzwords. These are all fads as fleeting as they are annoying, and the enemy of clear communication. Chapters 3 and 4 will allow us ample space to explore just how damaging they can be, and how we can resist them.

Perhaps it would be more useful for IT specialists to think of how to communicate the standards and underlying principles they use. How many non-IT specialists know about the NIST[2] and its Five Pillars, or COBIT[3]? These are sound, common-sense, principles-based assets to any organization. One thing IT specialists can do to bridge gaps is to consider ways to educate non-IT teams in how they see the world. What current and emerging risks concern them? These will often be easy to explain in plain language. How well does the organization's IT governance support its risk management? Which IT projects should the organization prioritize? How should it manage those projects and report on them?

This is, in an era of ever-increasing cyber-threats, crucial to any organization. IT specialists are indispensable to keeping their organizations safe, but they cannot do so alone. As the European Cyber Security Organisation's [ECSO's] 2021 CISO [Chief Information Security Officer] Survey Analysis Report says, "It is commonly agreed on that in order to align cybersecurity with business priorities **CISOs not only need to manage up to their Boards, but also manage across to the employees/rest of the company or organisation.**" [bold in original][4]

Few would disagree with this statement. After all, there is little point in hiring the best IT specialists and then ignoring their advice. Nor should the

IT specialists do their work in a vacuum; as with any part of the organization, IT must align its objectives and priorities to the organization's. Whatever an organization's risk threshold, in most cases it makes sense to adhere to it. (Obviously, this excludes criminal activity or indeed anything that will actively damage the organization.)

This assumes that the board and senior management have agreed, articulated, and clearly communicated both strategic objectives and risk appetite. Whether they have done so, and how they have done it, reveals a great deal about the organization's culture. It can be difficult to educate colleagues and improve controls under the best conditions; people are busy with their "real" work. It's even harder if the message everyone receives is inconsistent and weak. People may think that IT security – or indeed any risk management activity – is unimportant, or concerns only those working in IT security, risk, audit, or compliance.

The ECSO rightly points out that

> Company culture and evolution of mentalities remain extremely slow, and it is still a long road. CISOs are often met with resistance when trying to implement a cybersecure culture in their companies and push employees to have a more cyber-hygienic workplace environment.[5]

Anyone can see that company culture will either encourage or hinder IT security efforts. One common obstacle to a good IT security culture is the language used. However, the problem is not the technical terminology most IT specialists reproach themselves for using. Instead, it's what organizations do more broadly – using long words and too many of them, overcomplicating the simple.

Looking at the quote above, the ECSO wants companies to have "a more cyber-hygienic workplace environment." This is both wordy and vague. How far does one go in creating an entire "workplace environment" (as distinct from simply a "workplace")? Who came up with "cyber-hygienic"? For many people, this wording may create an image of robots moving around a stainless-steel, minimalist space. No wonder they don't see how it applies to them.

Yet it does, and it's as simple as saying things like,

> Don't share your user name and password with colleagues – and definitely don't stick them to your computer screen! Don't give team members access to databases they don't need for their jobs. Make sure you cross-shred sensitive documents *before* the cleaners come in. Lock away your files when you're not at your desk.

After all, most data breaches come from inside the company – not from sinister external forces.[6] Nor are all these breaches linked to deliberate malfeasance. Often it's a simple oversight, someone failing to follow a procedure or not even knowing about it.

Many people assume that it takes IT specialists to understand IT security; however, often it's the most basic controls that make the greatest difference to an organization's security culture. After all, how many people claim they can't be expected to understand physical access controls because they're not locksmiths or CCTV installers?

All of this seems obvious enough. However, organizational culture often discourages or even blocks communication among teams. John Chesshire, quoted in Chapter 1, says that the single biggest barrier to effective communications is

> In many ways the culture of the organization. I have worked in one organization in particular where communications between functions were inexplicably poor, for no obvious reason. After much consideration, I can only put it down to the culture of the place and some of its leading personalities. A sense that somehow, knowledge is power, knowledge sharing and comms are unnecessary, and constructive dialogue is a waste of one's time. I have never subscribed to these viewpoints, by the way!

Given how senior Chesshire's position was in the organization in question, this leads us back to "tone at the top." If "leading personalities" permit or even encourage a counterproductive culture, it can be hard for even senior colleagues to change things. Boards and senior managers are responsible for creating constructive, positive cultures in their organizations, and making sure that this message reaches *everyone*. If this message permeates the culture of the organization, then not only will people communicate more effectively across the organization, they will also do so with those at the top of and outside the organization.

UPWARD AND OUTWARD

You can probably imagine dozens of ways to share everyday, common-sense advice that many people overlook, or don't realize is important. You don't need to understand technical terminology, or server architecture, or binary code – just plain language and a desire to share useful information.

Using plain language to communicate information is critical to another group often seen as a distinct species within organizations: executives. Many clients will say to me, "But you don't understand – we need to communicate at the highest levels!" Actually, I do understand; I worked directly for the group CEO of a global bank for several years. What he expected – clear, concise, factual information delivered as quickly as possible – was the opposite of what many people think CEOs want. Which leads me to wonder…are CEOs aliens?

Well, clearly many people think they have unearthly powers. Yet few stop to think that CEOs are busy people; as André Spicer says, "Executive attention is one of the most scarce resources within any large organisation."[7]

Executive attention, though scarce, is often abused. Jane Bettany says, "I've seen board packs that are hundreds of pages long which seems unreasonable to me. I saw one CEO pile up the executive/senior management team papers for six months, and the pile was taller than he was. Is that good use of a CEO's time?"

The answer is clearly no – but where does this phenomenon come from?

Often it arises from people's desire to show how much they know and have done. Many people feel insecure in their jobs and thus take any opportunity to demonstrate their value to the organization. Ola Bello feels this is human behavior common to all organizations: "Irrespective of size, senior management and executives are constantly balancing how they are perceived, limited resources, proportionality, materiality, and the bottom line. I think what people have to lose gets in the way of effective communication mostly."

I've seen 300–400-page reports – often in a font so small no person can reasonably read it – go to the people at the highest level. At least 90% of what is in those reports is either completely irrelevant or features a level of detail inappropriate for the board. The 10% that is material is often so badly written that readers cannot extract anything meaningful from it. Clearly, the cause of this is cultural. If people feel they have to waste other people's time in a vain attempt to shore up their own job security, then something's wrong in the organization.

Let's now look at reporting outside the organization. Chapter 3 will feature case studies, but it's worth spending time now seeing how reports with a low content-to-length ratio can affect not just an organization, but an entire sector.

Since the financial crisis of 2008, regulators have increased scrutiny (or the appearance of it, at least). Many financial services organizations have responded by increasing monitoring, leading to more reporting and bigger report-producing teams. Yet has any of this improved matters?

Consider what central banks have said in recent years about resilience in their respective systems. In December 2019, the Bank of England published its report on the results of stress testing in UK banks. The report stated,

> The 2019 annual cyclical scenario stress test (ACS) shows the UK banking system would be resilient to deep simultaneous recessions in the UK and global economies that are more severe overall than the global financial crisis, combined with large falls in asset prices and a separate stress of misconduct costs. It would therefore be able to continue to meet credit demand from UK households and businesses even in the unlikely event of these highly adverse conditions.[8]

This sounds extremely encouraging; however, the Bank canceled the exercise in 2020 because of the Covid-19 pandemic.[9] Only four months earlier it had expressed confidence that the banking system would cope with "deep simultaneous recessions in the UK and global economies that are more severe overall than the global financial crisis" – in other words, exactly what happened as a result of the pandemic. The 2019 report, incidentally, did not even mention the possibility of a pandemic.

The European Banking Authority also delayed its annual stress-testing exercise by a year in response to the pandemic.[10] However, the European Central Bank and the US Federal Reserve and both published stress-test results reports in 2020[11], and included the Covid-19 pandemic in their considerations.

Why does this matter? It matters because the financial services sector – on which it's fair to say the world depends – responds to its regulators. Banks' reporting focuses on meeting regulatory standards, mirroring regulators' priorities and even language. But if a regulator itself isn't reliable, then banks are spending a huge amount of time and money on an empty exercise. What's more, the exercise itself produces variable and often questionable results, as Deloitte demonstrated in a review of Irish banks' regulatory reports.[12] In the meantime, what else could banks better spend their time doing?

Given how often banks – to take just one example – create a reporting industry, it's worth questioning the value of the work. Numerous entire teams devote every day to producing reports – daily, weekly, monthly, quarterly, and annual. Once they finish the cycle, they start again, and multiple layers of management have to review, wordsmith, and argue about every version. The content of many reports overlaps, so there is not only duplication but also room for error in not carrying through every single change in one report to the others. It is no exaggeration to call this an industry.

The reports this industry produces go to executives and regulators, but for different reasons. Executives need reports to help them make decisions. Regulators need reports to check compliance. Therefore, many reports that go to both groups attempt to meet both their needs. This is difficult to achieve, and the result is often a lengthy report that doesn't clearly meet either group's objective easily. (Again, the Deloitte report indicates this.)

The risk to organizations from this situation is enormous. If reports don't convey clearly what regulators need to know, organizations risk increased scrutiny, fines, possible bad press coverage, and even the loss of their license to operate. If the reports don't give executives what they need to make decisions, this could affect the survival of the organization itself. This is no exaggeration – the sheer scale and impenetrability of many reports make this a certainty, as case studies in Chapter 3 will show.

Given that report-writing training is my livelihood, you may think I'm destroying my business model with this argument. But I'm not saying ban reports. I'm saying that what reporting teams currently produce often introduces one of the biggest sources of uncertainty into the process. What should be a control – providing reliable information in a reader-friendly form to aid regulatory compliance and senior decision-making – undermines itself. It increases, rather than mitigates, many serious inherent risks, such as financial loss due to human error in decision-making, or regulatory fines.

Our shared goal must be more and better communication throughout and beyond organizations. This includes fewer but more useful reports in any organization, no matter how small or large. Quality, not quantity, should be everyone's focus.

ACTIVITIES

- Map your own professional and team cultures to the analysis you did at the end of Chapter 1. How do your professional education and qualifications influence how you communicate? Do you assume your immediate colleagues share the same background and assumptions?
- Think of who – outside your team or department – you communicate with most often. Is it someone in a different part of your organization? A client? A regulator? How will their background and assumptions differ from yours?
- Look at an email, report, or other pieces of writing that you sent to this person. Did you receive the response you expected, wanted, or needed? If not, why not? Could different approaches to communication explain why?
- How do you communicate within your team and with other teams? Do face-to-face (including using videoconferencing) or telephone conversations dominate? Or do you rely on sending written communications back and forth, creating an audit trail?
- How has your communication changed in recent years? Has working from home or remote working affected your working relationships and reporting?
- How does your organizational culture affect executive and regulatory reporting? Do you think there's a tendency to include too much information? How much attention do your reporting teams or colleagues pay to different audiences' needs?

SUMMARY

- We have looked further at cultural and organizational factors in communication. What may seem obvious can change drastically depending on the reader's or listener's background, context, or objectives.
- Just because people work in the same organization doesn't mean that they share the same assumptions, values, and objectives.
- Given how many waking hours we spend working, it is logical to assume that our teammates share our views and priorities. But sometimes the familiarity we think will create the perfect conditions for effective communication backfires.
- We all need to be aware of the gap between one's professional group culture and others, without overstating differences. IT specialists are often the most conscious of the need to adapt their terminology and approach to non-specialists.
- Communication isn't only problematic within and among teams. Some of the most business-critical writing goes to executives and regulators, for different reasons. However, if reports don't convey accurately and clearly what executives or regulators need to know, organizations risk serious consequences, as Chapter 3 will show.

NOTES

1 Margaret Heffernan, *Wilful Blindness: Why We Ignore the Obvious* (London: Simon & Schuster, 2019), 43.
2 U.S. Department of Commerce, National Institute of Standards and Technology, www.nist.gov/
3 ISACA, Cobit, www.isaca.org/resources/cobit
4 The European Cyber Security Organisation (ECSO) Users Committee, Survey Analysis Report: Chief Information Security Officers' (CISO) Challenges & Priorities, April, 2021, https://ecs-org.eu/documents/uploads/uc-ciso-survey-analysis-report.pdf, 49 (bold in original).
5 *Ibid.*, 49.
6 Marc van Zadelhoff, "The Biggest Cybersecurity Threats Are Inside Your Company," Harvard Business Review, September 19, 2016, https://hbr.org/2016/09/the-biggest-cybersecurity-threats-are-inside-your-company. For more recent insights, see UK Department for Digital, Culture, Media & Sport, "Cyber Security Breaches Survey 2021," March 24, 2021, Cyber Security Breaches Survey 2021 - GOV.UK (www.gov.uk).

7 Spicer, *Business Bullshit*, 133.
8 Bank of England Financial Policy Committee, Financial Stability Report, December, 2019, www.bankofengland.co.uk/-/media/boe/files/financial-stability-report/2019/december-2019.pdf?la=en&hash=99431A541357AC6D601A99B9 50455E2344C12901
9 Kalyeena Makortoff, "Bank of England cancels stress tests for UK's biggest lenders," The Guardian, March 20, 2020, www.theguardian.com/business/2020/mar/20/bank-of-england-cancels-stress-tests-for-uk-biggest-lenders-coronavirus
10 European Banking Authority, "EBA updates on 2021 EU-wide stress test timeline, sample and potential future changes to its framework," July 30, 2020, www.eba.europa.eu/eba-updates-2021-eu-wide-stress-test-timeline-sample-and-potential-future-changes-its-framework
11 European Central Bank, "Euro area banking sector resilient to stress caused by coronavirus, ECB analysis shows," July 28, 2020, www.bankingsupervision.europa.eu/press/pr/date/2020/html/ssm.pr200728~7df9502348.en.html; Board of Governors of the Federal Reserve System, "Federal Reserve Board releases results of stress tests for 2020 and additional sensitivity analyses conducted in light of the coronavirus event," June 25, 2020, www.federalreserve.gov/newsevents/pressreleases/bcreg20201218b.htm
12 Deloitte, "Solvency and Financial Condition Report (SFCR): Spotlight on year one disclosures," www2.deloitte.com/ie/en/pages/financial-services/articles/solvency-and-financial-condition-report.html

Chapter 3

Clarity
The theory

The first two chapters have discussed the unconscious and subconscious factors affecting how we communicate. These include everything from national culture to team-specific specialist training, all of which conditions us to see the world and articulate it in a particular way. Unsurprisingly, our reports will reflect this worldview, sometimes to the detriment of clear communication.

This chapter takes us further into the realm of conscious and unconscious thought. However, now you will see even more examples that relate directly to your everyday writing. By bridging the gap between our assumptions and words, we can produce clearer, better, more useful writing.

The first principle is that clear thinking and clear writing are linked. If you are unclear about your own thoughts, how can you possibly articulate them clearly? This may seem obvious, but under the pressure of reporting deadlines and corporate expectations, clarity is often a casualty.

This insight is not new. In his 1946 article "Politics and the English Language," George Orwell discussed many of the bad habits humans fall into when trying to communicate. What is more useful, though, is his point that humans often deliberately exploit bad habits to communicate badly – to obscure, mislead or simply lie. He called these a "catalogue of swindles and perversions"[1] – swindles and perversions many of you will recognize as standard in your workplace.

Words matter. We know this, which is why, depending on our cultural conditioning, we favor or shy away from certain terms. We know that they are likely to produce a particular response, whether good or bad. So far, so normal – this is what communication should be about. We think of what we have to say and why, and how we want our reader or listener to act; we then choose the words likeliest to prompt that result.

How often, though, do we do the exact opposite? Instead of choosing our words to produce a particular effect on the reader, and resulting action, we speak or write as if on autopilot. Then we wonder why people yawn in meetings, furrow their brow in confusion over reports, and fail to take the action we expect. This, unsurprisingly, is the opposite of what clarity should achieve.

DOI: 10.1201/9781003422365-4

We all recognize the truth of what I've just described; nevertheless, we persist. Why? I suggest there are several underlying, very human, factors: habit, tiredness, and fear.

HABIT

Habit takes us back, in part, to the previous two chapters. Whether we speak about cultural conditioning or corporate expectations, we all create habits that, for the most part, serve us well. If we are speaking to someone from our own culture, or from another one we understand well (or even better, speak the language of), we use shortcuts in wording and reasoning, confident the other person will understand. Similarly, if a cybersecurity expert meets a peer at a specialist conference, they can happily discuss the latest trends using the technical terms common in their field. Where there is shared understanding, these habits are effective and efficient.

When, though, do they become counterproductive? First, when the habits are those of using terms carelessly. Carelessly can mean without considering the actual meaning; it can also mean using a term to mean multiple things, without clarifying one's own meaning. One excellent example, which I will carry through this chapter, is "issue."

What do people mean when they use the word "issue"? Without context, it's difficult to know; it can mean a problem, an audit or other finding, a topic of conversation – even children. ("He died without issue in 1756.")

Most reports I have read – and at the time of writing, I estimate these to be well over 3,500 – use the word "issue" to mean both "finding" and "problem." Now, in audit, risk, and compliance reports, an "issue" meaning a "finding" will usually report a problem. Someone isn't adhering to certain controls, so the organization is at risk. Some people also use the term "issue" to mean a risk that has materialized, moving from something that may happen to something that has happened. (I prefer to call this kind of situation an "event" or "incident.") So, even within a specific professional context – audit, risk, and compliance – there is room for misunderstanding.

In other settings, though, it can be even more unclear. Imagine an internal memo in which the writer mentions "issues around resource." Leaving aside the now-common all-purpose use of the preposition "around" to mean "about," "with," "on," and so on, this sentence is unclear. What does the writer mean? I have seen this exact wording in dozens of reports, and not once has it been clear what it signifies exactly. Does it mean that there aren't enough members of staff? That there are enough, but they haven't received training? Or that there are enough, all fully trained – but they still can't do what they should? Each interpretation gives us a different problem to solve: either hiring more people, training those we have hired, or going back to the recruitment and training process to see which flaws in the process have created the situation.

Clearly, phrases such as "issues around resource" don't inform the reader. They're wasted opportunities to communicate specific information and prompt reflection or action. What they are is a habit – exactly what Orwell described when he said that in modern English, "prose consists less and less of words chosen for the sake of their meaning, and more and more of phrases tacked together like the sections of a prefabricated hen-house."[2]

Using language in a clear, fresh, useful way is a hard habit to develop and requires going against what may be corporate habit and culture. By questioning the words you and others are using, you may be questioning assumptions and even people's sense of belonging. After all, using the same language as everyone else makes you part of the tribe.

> Some psychologists point out that we are mental misers: we often prefer the ease of hanging on to old views over the difficulty of grappling with new ones. Yet there are also deeper forces behind our resistance to rethinking. Questioning ourselves makes the world more unpredictable. It requires us to admit the facts may have changed, that what was once right may now be wrong. Reconsidering something we believe deeply can threaten our identities, making it feel as if we're losing a part of ourselves.[3]

I'm sure you can all think of times when you have, out of habit or under pressure, created a report by selecting a standard phrase, adding a sentence your supervisor approved in a previous report, and copied boilerplate wording from your organization's intranet. You probably knew at the time it was a shortcut, and not an effective one, either. But you were under pressure to finish drafting the report, and these wordings were already available, so why not use them? After all, no one can then blame you – they're clearly acceptable in your workplace, so you risk nothing by using them.

Nothing, that is, except clarity and meaningful action.

We may think of habits as things we actively cultivate, but the linguistic habits we're discussing here often make their way into our brains unbidden. The novelist David Foster Wallace, discussing language with the lawyer and language expert Bryan A. Garner, described the process perfectly.

> On another level, when a vogue word – to dialogue, to proactively dialogue, or something – when it enters the mainstream, I think it becomes trendy because a great deal of listening, talking, and writing for many people takes place below the level of consciousness. It happens very fast. They don't pay very much attention, and they've heard it a lot. It kind of enters into the nervous system. They get the idea, without it ever being conscious, that this is the good, current, credible way to say this, and they spout it back.[4]

If this sounds familiar, don't beat yourself up about it – well, not too much. Even Orwell recognized the all-too-human instincts behind this writing-by-numbers approach: "The attraction of this way of writing is that it is easy."[5] It is easier – it's quicker, it feels safer, and therefore it's the most tempting option when writing in a hurry. However, it's a wasted economy. If it allows you to produce that draft more quickly, what happens when your manager reviews it? If your organization is lucky, the manager will immediately recognize the clichéd shortcuts and will ask you to re-draft, to clarify.

If your organization is less lucky – and is typical of many organizations – the manager will be so used to seeing this writing style that he or she will not see the problem. It's possible that the manager will recognize the draft as being hard to read, but, being equally conditioned by the organization's reliance on unclear writing, will try to fix the problem by adding...more unclear writing. Chapter 9 will discuss the reviewing process and how often both ego and habit conspire to make it counterproductive. (It will also advise you on how to improve the process.)

However, the likeliest situation is that the manager will not read the draft closely – it's too vague and impenetrable for that – but will assume it means what he or she thinks it should mean. It then goes up the management chain, however long or short, with each "reviewer" spending valuable time reading the unreadable, adding some further wording to put a personal stamp on it and show they've done their job, and sending it on. By the time the final report goes out, it is a collection of various people's bad habits, personal quirks, and unconscious regurgitations of whatever verbiage is fashionable that week. It's certainly not a clear communication of important information that senior decision-makers need to take action.

This is clearly a waste of time and effort, but many organizations persist in writing without thinking of readers. As I've just explained, writing that taxes the readers' brains is writing that goes nowhere.

Writers specializing in the neurological impact of reading confirm this. Bill Birchard emphasizes what a relief it is for readers not to tax their brains. "People's brains work faster when processing less, whether it's processing words, pictures, objects, or ideas. They in turn love how you ease their mental load, or as scientists say, boost 'processing fluency.'"[6]

Boosting readers' "processing fluency" in turn improves decision-making. So, if your corporate habit is to produce wordy, convoluted writing, it almost guarantees two negative results. First, most readers won't read it, or may start but abandon it. Todd Rogers and Jessica Lasky-Fink discovered that,

> In a recent survey we conducted with around 1,800 working professionals, they estimated that they delete about half of the emails they receive without reading them. ...This is a striking result when you think

about it from a cognitive perspective. Busy readers routinely decide how valuable a message is *without actually reading it!*[7]

All this arises from habit. As I mentioned earlier, habits are human, and they can be good – effective, efficient, productive – as well as bad. However, we tend to fall back on bad habits when we are under pressure, stressed, and, above all, tired.

TIREDNESS

Don't underestimate tiredness. Many, if not most, people you interact with at work will be tired. Maybe they have too much work, or think they do; maybe they have problems at home, or are unwell. In any case, they will not have time, energy, or mental stamina to spare.

As writers, when we find ourselves tired for any reason, it is tempting to choose the easy option. Shortcuts save the very time, energy, and mental stamina that we lack. As Orwell wrote, this approach means that "you save much mental effort, at the cost of leaving your meaning vague, not only for your reader but for yourself."[8] After all, as Margaret Heffernan reminds us, "Too often, actually thinking about or reflecting on a decision that we face is far too taxing."[9] Working out what we mean and then choosing our language carefully to convey it is indeed taxing.

Furthermore, we make decisions in doing so: How exactly do I tell senior managers about their incredibly expensive IT project that's not going to work? Is it my place to do so? Can I refer vaguely to some superficial work done by our external auditors last year and avoid saying anything more specific? How will they react? Will it affect my job? Could I phrase things in a way that makes it look as though I've done my part – producing a report – while not saying anything that can come back to bite me?

These are all indeed taxing questions; they're unpleasant, and the answers available hardly less so. As with the manager reviewing an unclear report, you could ask hard questions and insist on clear answers. Or you could take the easier, more obvious, option and continue to express yourself vaguely. This is certainly a short-term option – but it leads to greater risk in the medium and long term, not only for you, but also for the organization.

It's hard to force yourself to avoid that risk. Orwell writes of the discipline needed to work out what we have to say, and the fewest, most precise words to say it; however, as we've all seen, we don't have to do this.

> But you are not obliged to go to all this trouble. You can shirk it by simply throwing your mind open and letting the ready-made phrases come crowding in. They will construct your sentences for you – even think your thoughts for you, to a certain extent – and at need they will perform the important service of partially concealing your meaning even from yourself.[10]

This raises the question of why would you want to conceal your meaning from yourself. I can think of several reasons. Perhaps you're too close to the material you're trying to articulate. You don't, with the best will in the world, have the necessary distance to see what is and isn't clear to someone else, without your expert knowledge.

Perhaps you are under pressure and falling back on the "safe," manager-approved linguistic habits of the team or organization.

Or maybe you're afraid of what will happen if you do tell those senior managers their IT project won't work.

FEAR

It can be difficult to work in audit, risk, compliance, and IT. Leaving aside specialist terminology, often what people in these areas are saying is that something has gone wrong. There is a problem, or there is likely to be a problem. There is a risk.

It is absolutely normal to feel apprehensive about delivering bad news; we wouldn't be human if we didn't feel this way at least sometimes. And, as we showed in Chapter 1, certain cultures are more actively evasive in the face of potential conflict. Whether it's down to cultural value in "saving face," or having grown up equating direct speech with aggression, it can be hard for people to convey unpleasant but necessary information.

Perhaps this is why the Institute of Internal Auditors has, in its new Global Internal Audit Standards, made the very first Standard (1.1) "Honesty and Professional Courage." Mindful of the pressures those in assurance functions can face – after all, we rarely deliver good news – the Standard charges the chief audit executive with leading and protecting the team. "The chief audit executive must maintain a work environment where internal auditors feel supported when expressing legitimate, evidence-based engagement results, whether favorable or unfavorable."

Organizations are no different, and the "tone at the top" discussed in Chapter 2 reveals what people seek, accept, or avoid. If a board or senior decision-making committee actively encourages dissenting views, extensive research, or plain language, it's likely their colleagues will be less fearful of delivering bad news. However, if the highest levels of the organization encourage groupthink, superficial reasoning, or evasive corporatespeak, fear will spread.

In *Wilful Blindness*, Margaret Heffernan expertly describes and dissects the all-too-human tendency to quash individual insights and behavior in favor of communal norms. This tendency arises from tactics necessary for survival, and many people would say it's still necessary to survive in their workplaces.

But what does that mean for the survival of the workplace – of the organization?

Wells Fargo gives an example of what happens when those who should lead – chief audit executives and chief risk officers, for example – don't. In a damning federal judgment, Judge Christopher B. McNeil fined both the chief audit executive and the chief risk officer of Wells Fargo for multiple failings, not least in misleading reports and other communications.

> Respondent McLinko [the Chief Audit Executive] failed to maintain independence and objectivity in his relations with the head of the Community Bank, Ms. Tolstedt, and the Community Bank's Group Risk Officer, Ms. Russ Anderson. This included his strong endorsements of Ms. Tolstedt's risk management practices notwithstanding his review of the independent audit report indicating that more than 1.4 million accounts had potentially been subject to simulated funding and not-withstanding his review of the 2016 report by the OCC examiners find-ing the Community Bank's sales practices risk management was unsafe or unsound. It also included his cloying praise of both Ms. Tolstedt and Ms. Russ Anderson in correspondence exchanged throughout 2013 to 2016, at a time when neither Ms. Tolstedt nor Ms. Russ Anderson were willing to acknowledge the root cause of team member misconduct.[11]

An internal investigation report from 2004 warned of the financial and reputational losses Wells Fargo would go on to suffer. "However," as Srinvisavan Ragothaman, Tyler Custis, and Melissa Christianson note in their analysis of the scandal, "this report failed to make an impact on senior managers. Additionally, internal auditors failed to continue to press this issue for the next several years as the fraud ran rampant."[12]

You may feel the Wells Fargo case is an extreme one – but organizations every day succeed or fail because of their communication skills.

CLARITY AND SUCCESS

"Sadly," writes André Spicer, "well-reasoned words are rare in corporate life."[13] Whether through habit, tiredness, or fear, people write a lot – and most of it is bad: wordy, unclear, often pretentious.

But why does it matter? It matters because unclear writing betrays either unclear thinking or the intent to deceive. It's worrying enough when people write badly to actively mislead the reader, or hide the unpalatable; worse yet when they do it unthinkingly, simply because this is what they've always done.

It matters because when people write badly, organizations suffer. They are inefficient internally because employees waste time and energy clarify-ing points. When people don't clarify matters, though, they will fail to take necessary action. Why would they? They didn't realize the report was telling them to.

Organizations also suffer externally. Ironically, most people who write badly in organizations quickly spot wordy, unclear, pretentious, or evasive language when they receive it. Give them a press release from the utility company that has just had to confess to overcharging, or from their favorite online retailer who has admitted exploiting workers. If the press release is in managementspeak, even those who inflict it on others will see the telltale signs of evasiveness and falsehood. Even companies who should know better resort to this almost instinctively: "The great enemy of clear language is insincerity. When there is a gap between one's real and one's declared aims, one turns as it were instinctively to long words and exhausted idioms, like a cuttlefish spurting out ink."[14]

One example of this is from the executive summary of the official report on BP's Deepwater Horizon accident investigation. The report itself is concise, clear, and straightforward, even for a layperson. However, the first page of text ends with these two sentences:

> Wherever appropriate, the report indicates the source or nature of the information on which analysis has been based or conclusions have been reached. Where such references would be overly repetitive or might otherwise confuse the presentation, evidentiary references have been omitted.[15]

What impression does this create? Would you have confidence in the ensuing executive summary and indeed report? Or would you feel the writers were seeking to evade and excuse? You wouldn't be alone in saying the latter.

And yet I have sympathy for the writers of those sentences. (I think we can assume the report was written by committee.) They knew the report wouldn't have good news, and they also knew that they didn't have all the answers they or the readers expected. In those circumstances, it's normal to write something alerting the reader to any gaps or limitations. How else could the writers have expressed this entirely reasonable point? They could have been more direct and simple, saying something like, "This report gives the source and nature of information wherever possible. Where we feel evidence is repetitive or potentially confusing, we have put it in the appendix."

Perhaps my re-write over-simplifies, but the effect is completely different from the original. It says that the writers trust the readers. Instead of a paternalistic and patronizing message ("We decide what's appropriate, and we know how easily confused you get"), it says "There's a lot of information here, so we've organized it in a way we hope is clear and useful to you."

We'll return to this example in Chapter 4, to see in more detail how structure and word choice create such different impressions in the two versions. However, the difference between the two versions can be the difference between failure and success in handling a disaster.

BP's share price plummeted after the accident, but better communication could have mitigated this. It seems obvious, yet many organizations persist in failing to make the connection between clear writing in plain language and good performance. There is indeed one, as author and writing expert Rupert Morris demonstrated nearly 20 years ago, with his Clarity Index.[16]

This simple graph showed a compelling correlation between clarity and performance. Morris and his colleagues studied the annual reports of FTSE 100 companies and analyzed them for clarity: simple, short words and sentences, favoring the active voice.[17] (Chapter 4 will go into these attributes in more detail.)

They then tracked the organizations with the clearest writing against average stock-market performance 2002–2003. As you can see, the top line shows that those organizations with clear, simple, reader-friendly reports consistently outperformed the FTSE 100 and therefore their peers.

The Clarity Index - last 12 months

Another positive example comes from the US, from Berkshire Hathaway, the investment company chaired by Warren E. Buffett. Few would dispute either the company's or its leader's success. Is there a connection between this success and clarity? I think so.

Berkshire Hathaway's annual reports are in the public domain, and the most revealing part of them is Buffett's annual letter to shareholders.

One of my favorite examples of Buffett's plain-language style comes from his 2012 letter. In it, on page 1, he starts with a brief paragraph (80 words across four sentences) setting out headline figures.

He then writes, "A number of good things happened at Berkshire last year, but let's first get the bad news out of the way." He provides a bulleted list of "failures" that take up the rest of the page, with the last point on the page being, "The second disappointment in 2012 was my inability to make a major acquisition. I pursued a couple of elephants, but came up empty-handed." It's not until a third of the way down the next page that he says, "Now to some good news from 2012[.]"[18]

You could say that his style is folksy, and that that won't go down well in your organization; you may well be right. The point is that wherever you look in his letters to shareholders, from whichever year, you will find plain language. There is no attempt to evade, obscure, or sugarcoat.

What's more, throughout his letter, he demonstrates true executive accountability. When something goes badly, he uses the word "I." When things go to plan, he says "we." And when he wants to praise a particular aspect of the company's performance, he names others – not himself – as deserving the credit. He credits them by name, and says "they." This is far removed from many people's experience of managerial accountability, which sadly often sees the most senior people claiming the rewards for success, while blaming others when things go wrong.

In Buffett's letters, we see the confidence of a person and organization prepared to say openly what they have done, why, and how, whether it has succeeded or failed. When the confidence exists to communicate clearly within an organization, it inspires confidence outside an organization.

LACK OF CLARITY – AND FAILURE

Let's consider now some counter-examples, where the "tone at the top" discouraged plain language and clear writing. One obvious one would be practically anything from the UK banking system leading up to the 2008 financial crisis. Michael Lewis has written compellingly about the systemic failures and groupthink that allowed the crisis to develop and failed to mitigate its risks.[19]

However, a BBC News online article from the year of the crisis summarized handily the way in which reporting language enabled problems to continue and worsen.[20] In it, a risk manager from HBOS (one of the two UK government-rescued banks) tells the reporter that "We were constantly being pushed back and constantly being told that things had to be put in a way that wasn't going to upset the business – basically that wasn't going to stand in the way of the sales culture at the time." Who was "pushing back" and telling people not to "upset the business"? The head of risk, who led the function.

Organizational leaders generally *choose* how to communicate; they know perfectly well the difference between clear and unclear writing. André Spicer takes Nokia as an example, usefully contrasting clear, plain communication (the CEO's famous "burning platform" email) with turgid, evasive, corporate jargon-laden memos about looming redundancies. He also describes the culture of fear within the organization, discouraging clear communication and ultimately truth-telling.[21]

Another corporate example of bad practice was GM, which Margaret Heffernan discusses at length in *Wilful Blindness*. Not only did the culture at GM discourage clear communication, it ran internal training on how to avoid it. In his report to GM management about ignition switch recalls, Anton R. Valukas refers to "tone at the top" and corporate culture. In the section of his report entitled "Resistance to Raising Issues" (and "issues" is a gentle term, remember), he describes an atmosphere of not simply resistance but outright repression. Not only did employees trying to report problems encounter "pushback," they were told how to express themselves.

"Employees were given a number of words to avoid, with suggested replacements," Valukas writes. Instead of "Problem," they could use "Issue, Condition, Matter." Similarly, instead of reporting a defect, staff members were told to say, "Does not perform to design." [22]

I'm sure many of you can think of similar examples you've encountered in the workplace, hopefully with less severe consequences. Empty slogans, lazy buzzwords, and positive messages at all costs are a misuse of language and can lead to disaster.

Other corporate scandals show how difficult it can be for those supposed to advise organizations to do so. If there is a resistance, either within the organization or from major clients, to receiving necessary but unpleasant news, it's bound to compound any eventual problems. In the world of audit, external auditors can find themselves in a difficult position. Many bright graduates are eager to join household-name companies (many of which have been implicated in various accountancy scandals, from Greece's entry to the Eurozone to Tesco's misstatement in its accounts in the UK). Once there, they may find themselves working on a major client account, providing assurance over key financial controls.

There is, however, a potential conflict: if the client doesn't want to hear, for example, that its own accounting procedures are poor, what is an ambitious young external auditor to do? The big-name accounting firm may rely on that client, and a rolling contract for services, for both income and credibility. In such circumstances, it's not impossible for a junior external auditor to be encouraged to quash findings to please the client.

I know of several who moved into internal audit so as to retain objectivity and independence. However, internal auditors are also subject to pressure of a different kind, as internal politics play out through reports. In the world of audit, both external and internal auditors do the best they can under different circumstances. After all, their roles and skills, while often overlapping, *are* distinct. But it is fundamentally the corporate cultures they must work with that dictate how clearly they are allowed to communicate, and therefore how well they can do their jobs.

Mark Carawan, quoted in Chapter 1 on the common ground between internal and external audit, has seen how topical corporate culture has become:

> Ultimately, people's behaviors shape an organization's culture, and people change as individuals themselves, as well as there being a constant flow of individuals into, around, and out of an organization – nothing and no one is static. There has been much written about how incentive structures (monetary and others) are a major factor in driving behaviors, how leaders create "tone from the top" and cast the "shadow of the leader", how the middle management "marzipan" or "permafrost" layer is resistant to change, and other aspects of organizational behavior.

So far, we have looked at cases where those at the top have deliberately chosen to use vague or misleading language to obscure the truth. It can also happen inadvertently, often because of cultural differences, but also out of habit and corporate conditioning.

One good example of this comes – again – from financial services. You may remember the LIBOR scandal of 2012, when we learned that traders had been fixing overnight currency rates for years. As part of the investigations, emails among senior people in US and UK banking were made public, and the UK Parliament issued a report.

This report includes transcripts of interviews, including with the then-chairman of Barclays, Marcus Agius. When asked why he hadn't communicated the regulator's "concerns" to a new appointee, Agius gave a perfect example of corporate semantics and the wriggle room they create. "I would challenge the word 'concerns'," he stated. "That letter raises four issues and they are called 'issues'. The word 'concerns' I do not believe appears. I am not being pedantic but there is a difference between 'these are issues which I would like to raise with you' and 'concerns', which means 'I'm worried.'"[23]

The report's commentary on this exchange is restrained – perhaps too much so: "Barclays appears to have regarded the points raised by Mr Sants as 'issues' rather than 'concerns'. On the basis of the evidence it is unclear whether Barclays 'got the message'."[24] In saying this, the report allows this language to stand. It's possible that Barclays feigned innocent bewilderment at the use of "issues" rather than "concerns." It's also possible Barclays was genuinely confused – that the bank's hierarchy of corporate euphemism at the time did indeed rate "issues" lower than "concerns." Compounding all of this is the report's typically British coyness, stating that things are "unclear" – there's simply been a misunderstanding. Move along – nothing to see here.

Perhaps those in the UK banking system would have acted more quickly and decisively had anyone said, at any point, "This is fraud. It's a problem."

Then again, had anyone done so, the others would have had no excuse for taking action. This perhaps explains why vague, misleading language is so useful. It creates the perfect environment to avoid the unpleasant business of delivering bad news, establishing responsibility, and accepting criticism.

Even when criticism is clear, it can be difficult for organizations to turn it into meaningful and positive change. After the financial crisis, the UK Parliament investigated the role of the then-regulator, the Financial Services Authority (FSA). The resulting report stated – in a refreshingly direct way – that "By any measure the FSA has failed dreadfully in its supervision of the banking sector," which is clear criticism by any standard.[25] One would expect major reform after such stinging words.

The FSA did indeed change: by 2013, it had split into two bodies, the Financial Conduct Authority (FCA) and the Prudential Regulation Authority (PRA). However, did the culture change? If we gauge an organization's culture by its ability to communicate clearly, the evidence is mixed. As recently as 2019, the FCA produced a 77-word sentence as part of a final notice against a bank it had investigated and found against:

> On the basis of the facts and matters described below, SCB breached Regulations 14(3), 15(1) and 20(1), and failed to comply with Regulations 7(1) to (3), 8(1) and (3), and 14(4) of the Money Laundering Regulations 2007 (the "ML Regulations") by failing to establish and maintain risk-sensitive policies and procedures, and failing to require its non-EEA branches and subsidiaries to apply UK-equivalent anti-money laundering and counter terrorist financing ("AML") standards regarding customer Due Diligence and ongoing monitoring.[26]

So a full ten years after the House of Commons' clear criticism, and six years after a costly and complex restructuring, the UK regulator was still producing reports that didn't communicate effectively. Keeping in mind that this regulator publishes all its final notices, it's reasonable to expect a degree of clarity and conciseness. After all, how can the taxpayer and other interested parties keep abreast of what regulators and other enforcement agencies are doing, if they cannot read their work easily?

It's telling that in the early 2000s, I used excerpts from Enron's final letter to shareholders[27] during training. It was as an example of corporate waffle – throwing sand in readers' eyes, or squirting out cuttlefish ink to blind them. Everyone agreed, possibly with the benefit of hindsight, that the language was clearly designed to intimidate and impress, but not to inform. A few years later, around the time of the financial crisis, I saw people thinking the best way to safeguard their jobs, in times of mass redundancies, was to increase the verbiage in their writing. So I changed my approach – I would show clients an excerpt from Enron and an excerpt from their own writing, and ask which they found clearer. They were horrified to see that their own work was no clearer than Enron's.

Now, in many organizations across the globe, I meet clients whose work makes Enron's look like a model of clarity and transparency. While it's encouraging to know that people recognize they need help to write better, it's worrying that increasing job insecurity and non-stop work pressure make them likelier to write badly to begin with.

Remember the example I gave in Chapter 1, where the risk team hid the fact of fraud so as not to upset the managers committing it? Recall the example in this chapter, where the BBC reported on how HBOS' risk function diluted its reports to please senior staff? If you find yourself told to write reports to spare senior managers' or clients' feelings, you should ask why.

Obviously, if your reports include personal attacks, maybe your manager has a point – there's a reason why we often ascribe problems (not issues!) to "lack of management awareness" or "lack of diligence." It's more acceptable than saying "The managers are stupid and lazy," after all.

However, if your managers are encouraging you to avoid using clear language to say that there are problems, and what they are, when you have compelling evidence for it...maybe there are further problems. Maybe your organization has a culture that prefers to hide problems behind vague language. If so, the organization could face far more serious problems than a few senior managers' hurt feelings.

EGO AND OVERCOMPENSATION

André Spicer writes compellingly about how badly organizations encourage people to express themselves. The fear factor I've mentioned explains why many people's egos are bound up with how they write. It goes beyond cultural and profession-specific elements discussed in Chapters 1 and 2. It's more about how they position themselves in the organization, how they safeguard their perceived status or even their job. Never underestimate how many people fear losing their work, salary, and identity.

This means that every report becomes an opportunity to showcase the writer's (or writers') knowledge, experience, and hard work. There's nothing wrong with using all of these to provide information to people. The problem arises when the information comes second to the exercise of "showing off." This can take the form of using long words and complicated sentences ("Look how clever I am!"),[28] referring to true but irrelevant side-topics ("I'm not just a cybersecurity analyst – I also know about juggling!"), or including minute detail ("Look how much research I did!").

Weaning people off this habit can be difficult. They fear that without the verbiage, people will lose respect for them. This overlooks two crucial facts: the verbiage is stopping people from reading the text, and anyone who does read is likely to think less of the author for making their life so difficult. As

Bill Birchard advises, "Don't worry that simplicity will dumb down your writing. Great writers have shown for centuries that plain words serve just fine even for cosmic concepts."[29]

What, then, instead of plain words, tends to fill most pages? We could call it managementspeak, corporatespeak, business buzzwords, or sometimes something even ruder (as André Spicer does). It's – as we will see in Chapter 4 – the limited but empty wording people reach for simply to fill air or space. As Orwell said, this kind of verbiage serves two purposes: to speak or write without having to think, and to confuse. Many people will admit to the first one; few to the second. After all, the second implies a deliberate wish to deceive, which in turn implies incompetence or wrongdoing. No wonder few will admit to it.

We can add a third purpose, which is ego and overcompensation, both of which take us back to fear. Fear of not fitting in; fear of not being valued or respected; fear of losing status. All these fears operate in the workplace. We can all think of colleagues, managers, and clients who exhibit classic defensive or even aggressive behavior to compensate for these and other fears.

A more common and often acceptable way of managing these fears is to use managementspeak. It's an easy way to show belonging to a group. For junior members of staff, parroting the latest fashionable phrase, no matter how confusing or meaningless, reassures them that they are showing their commitment to the team. For those eager to climb the corporate ladder, speaking in almost entirely content-free sentences may seem a sure way to demonstrate their managerial skills. The irony is that many people at the top of the corporate ladder prefer short, clear communication. As one CEO told me, "If you can't tell me what's happening in half a sheet of paper, then either you don't know, or you're trying to hide something." This may seem harsh, but I've discovered it's often true.

Equally often, it's simply a case of misunderstanding what good communication means. Conditioned by corporate habit, many people lose the ability to talk to others in simple, human terms. I remember one colleague whose weekly updates were utterly impenetrable. One day, the manager told me and others that our confusion and boredom were becoming apparent during this colleague's updates. Why not, he suggested, focus closely on what the colleague was saying so that we could ask specific questions when he finished?

Let's leave aside the question of why the manager didn't encourage the colleague to express himself more clearly. My assumption was that he had, but it hadn't worked. Instead, we all strained our brains as this colleague proudly stated that his team had "leveraged blue-sky thinking going forward on an ongoing basis to benchmark continuous improvement in a world-class operational environment." We went quiet, then politely but nervously said, "OK – that sounds impressive. What exactly has the team done?" The colleague in question was at first self-conscious as he struggled for words. We had asked him to abandon managementspeak and tell us, in plain language, what had happened. It was an uncomfortable moment for him and for us, as were the next few meetings.

We continued to ask him direct questions and to encourage him to give direct answers. Finally, he came out with answers such as, "My team looked at what peer banks are doing and created a list of good ideas we can use." That was something we could all understand and congratulate him and the team for. It was clear that they had done something we would be able to use. However, it took weeks to get this colleague to leave his linguistic comfort zone and start communicating clearly. He wasn't and isn't alone: as André Spicer has pointed out, managementspeak can reassure insecure middle managers that they are doing the right thing.[30]

This is typical of many people's experiences. They listen to, or read, endless verbiage that says little. It takes time, energy, and, as my manager insisted, goodwill to encourage people to express themselves in listener- and reader-friendly ways. Often entire teams and organizations will tolerate the wordiness and vagueness in the belief that it's mostly accurate. This is a dangerous assumption, as wordiness and vagueness can often hide gaps, flaws, and misconduct.

Even if this is not so, wordiness and vagueness hide important information. Remember Spicer's comment about executive attention being a scarce resource.[21] People often feel inclined to over-write to impress senior managers, to justify their existence, or to anticipate all questions. The result is usually counterproductive, with the recipients losing both patience and confidence. Jane Bettany describes the vicious circle perfectly:

> When staff feel disempowered or low in confidence they tend to try to write more to demonstrate their worth and how hard they have been working rather than understanding the need to be succinct and focus on what you need from the person you are communicating to. The message can be lost that way.

Executives and senior decision-makers receive hundreds of pages of reporting regularly – a direct result of their colleagues' compulsion to over-write. The risks from this kind of excessive yet usually vacuous reporting are numerous. It's clearly inefficient, as Molly Young has pointed out: "No matter where I've worked, it has always been obvious that if everyone agreed to use language in the way that it is normally used, which is to communicate, the workday would be two hours shorter."[32] Most people would agree with this.

Then there is quality of work. How can senior decision-makers such as boards and committees make any decisions if they're relying on hundreds of pages of impenetrable verbiage to inform them? It's simply not reasonable to expect this even of well-rested, highly focused people. Their attention spans will wander; they will become distracted by true but irrelevant detail;

they will spot irritating typos, almost inevitable when a reporting team is producing too many words in too little time. The result is almost certain to be a vague impression that things are generally OK, with a few exceptions that (everyone hopes) are under control. After all, what does "issues around resource" mean? If you are seeing this as one of countless vague phrases across 400 pages, it's unlikely you will have the time, energy, or concentration to question one instance of unclear writing, let alone thousands.

In the years following the financial crisis, many people rightly criticized the chief executives and chairmen of banks involved. André Spicer refers to how often those questioned by government committees claimed they couldn't remember or didn't know about seemingly important topics.[33] On the one hand, I have some sympathy. I've reviewed 400-page packs of reports for bank boards and despaired, while trying to stave off the looming migraine prompted by such a mass of meaninglessness.

But I've also wondered how things could get to that point. CEOs may speak plainly and encourage it, but if the message gets lost further down, the reports will reflect this. Without emphasizing the need for concise, clear, plain language, the problem will continue and risk overwhelming even those at the top.

THE ETHICS OF CLEAR COMMUNICATION

"When possible, refer all matters to committees, for 'further study and consideration'."
"Bring up irrelevant issues as frequently as possible."
"Give lengthy and incomprehensible explanations when questioned."

You may think these instructions are from a tongue-in-cheek guide for new managers in your, or any, organization. Actually, they come from a document entitled "Simple Sabotage Field Manual," created by the Office of Strategic Services (OSS, the precursor to the CIA) in 1944.[34] How can instructions designed to undermine enemy forces have so much in common with many organizations' everyday habits? They clearly use humans' ability to complicate, confuse, and generally obstruct normal activity, but to different ends. In the case of the Field Manual, it's to win a war; in organizations, it's... what?

This chapter has discussed how people may unwittingly complicate language through habit, tiredness, or pressure to appear "managerial." It has also raised the fact that people in organizations may deliberately manipulate language to hide misconduct or evade responsibility.

This is why language has an ethical component – words matter. We know they can hurt feelings and even commit crimes. Issuing death threats is an illegal activity, yet relies solely on words. When organizations or government bodies talk about transparency, they are seeking to reassure the public of their trustworthiness. References to clear or plain language abound, to prove this commitment to transparency. What's more, the House of Commons' Public Administration Select Committee in the UK

> conclude that bad official language which results in tangible harm—such as preventing someone from receiving the benefits or services to which they are entitled—should be regarded as "maladministration". People should feel able to complain about cases of confusing or misleading language, as they would for any other type of poor administration.[35]

The US Government introduced its Plain Writing Act in 2010, and New Zealand a Plain Language Act in 2022. These are of course not the only countries to do so, and yet others feature plain-language versions of certain pages on their government websites. (Why all the pages of the government websites aren't in plain language by default is a valid question.)

It is not only public-sector-related bodies that may mandate plain language. After all, if plain language increases accessibility and understanding, why wouldn't every organization use it? Some, however, have to be forced to do so – in this case, by the regulator. As Bill Birchard points out,

> For many years, economists believed financial markets operated 'efficiently,' incorporating all available information. The research suggests otherwise. Markets incorporate all accessible information. No wonder that, in 1998, the US Securities and Exchange Commission mandated that companies use 'plain English' in reports to investors. The agency even published a 77-page handbook detailing time-tested tips – cut extra words, use active voice, speak with personal pronouns (e.g., you, your, yours), use common words for jargon, craft short sentences, keep subject, verb, and object together, and more.[36]

Many jurisdictions require clarity and transparency in consumer contracts, government documents, and other official forms. Enforcing it, however, is onerous, and it often falls to a consumer to complain that a merchant has unclear terms and conditions, for instance, violating the UK Consumer Rights Act 2015. Public bodies may say that they believe in plain language, transparency, and fairness, but do their actions bear this out? Consider the regulatory notice quoted earlier in this chapter; look up some of the guidance available in different countries on how to apply for visas or work permits. Given that many people relying on this guidance may not speak the language fluently, it's even more crucial that it be accurate, clear, and simple.

Rachel Browne, Audit Director at Audit Scotland, notes that "there's also consciousness that good governance is linked to openness and transparency. Internal communication in public bodies is often much more open than communication with the public." How well public bodies communicate can directly affect public trust. A government – or organization – that uses unclear language either doesn't realize it's failing to meet people's needs, or doesn't care. Neither situation will build, let alone maintain trust.

What is worse is when organizations unwittingly introduce terms that reveal a purely transactional view of people. We all know that private-sector companies seek a profit – they wouldn't survive otherwise. However, it is reasonable in most countries to expect a modicum of care for employees, customers, and communities.

What would you think of a company that refers to staff as "units of human capital" that are "terminated" when their careers in the company come to an end? I've seen this in a Big Four accountancy firm, a global bank, and a global manufacturer – no one can claim such clinical, even chilling language is sector-specific.

One company, writing about a project achieving its stated objective, claimed to have produced "the final solution." I had to explain that this term has a specific and horrifying meaning: the Nazis' plan to exterminate Jews. When I pointed this out to the company's employees, they were appalled – they immediately recognized the reference and how utterly unsuited it was to any context other than the historical. Yet someone, somewhere in this company had suddenly thought that in its polysyllabic finery, this term was a superior version of the old-fashioned, perfectly clear "stated objective." Others, hearing "final solution" in meetings, didn't question or contradict its use – they parroted it. This tone-deafness that so often leads to mere irritation in this case led an entire team to use Holocaust-specific terminology in an update to the board.

Words. Have. Meaning. When we lose sight of this, we not only risk confusing or even alienating readers; we risk our own ability to reason. Orwell made a compelling point about the habit of using a reduced, clichéd vocabulary, the kind so common in most corporate settings. "But if thought corrupts language, language can also corrupt thought. A bad usage can spread by tradition and imitation, even among people who should and do know better."[37] It's easier to mimic others' phrasing, thinking it cements our place in the team. However, it also saves us the trouble of thinking.

Then when we then *do* try to think independently, we find ourselves falling back on the limited, shopworn babble that passes for expression in many workplaces. Without exercising the mental muscle of our vocabulary, we end up leaning on verbal crutches either meaningless or unsuited to the task. Without an array of varied, nuanced terms to describe situations and propose responses, we cannot think rigorously or critically about anything. We will end up like the poor manager who couldn't express himself other than with "leveraged blue-sky thinking going forward on an ongoing basis to benchmark continuous improvement in a world-class operational environment." It took him a great deal of effort and discomfort to break the habit that had affected not only his communication but also his own ability to reason.

Those using unclear language don't only stultify themselves and bore their listeners; they spread the rot. As Molly Young writes, "The problem with these words isn't only their floating capacity to enrage but their contaminating quality. Once you hear a word, it's 'in' you. It has penetrated your ears and entered your brain, from which it can't be selectively removed."[38]

The next chapter will discuss in detail the hallmarks of unclear language in English and how to avoid them. It will use specific, report-related examples and encourage you to improve them. However, in order to benefit, you have to commit to the theory of clarity as well as the practice. The language we use is an ethical choice. Choose your words not only to achieve an operational end, but also to express a commitment to clarity and therefore transparency. If you do this, you will be truly radical!

ACTIVITIES

- How well do you think your organization communicates?
- What guidance or examples have you received on how to communicate? This can include official guidelines from the communications team, individual team standards, or even the tone expressed by the CEO in regular emails or intranet messages.
- What works well in your organization's communication and what doesn't?
- Look at the good and bad practices you've identified: what comes from "tone at the top," and what comes from habit?
- Does your organization use inclusive language? If not, what message do you think this sends to colleagues and clients?
- Have you tried to share, improve, or even create tools and techniques to communicate better within your organization? These can include checklists, examples in templates, or even discussing the topic during team meetings.
- Select two pieces of writing from your organization: one you feel communicates clearly, and one you feel doesn't. Make notes about how each makes you feel, and what you think contributes to that. It can be structure, language, or even presentation. You can then check your responses and develop them after reading the following chapters.

SUMMARY

- Writing in a clear style affects how well readers understand and then act on what they read. As a result, it affects how well organizations perform. This isn't solely about share price and other financial measures – it's also about the values underpinning the organization.
- It is hard to write clearly. Therefore, it's understandable that people often fail to, even when their job is to, produce meaningful reports for senior decision-makers. Habit and tiredness play a part; when people are under pressure and lack energy, they are likelier to reach for "safe," commonplace corporate expressions, even if these are vague or actively misleading.
- Ego and fear also play their part in this human drama. The larger the organization, the more hierarchical the culture, and the greater the opportunity and temptation to aspire for senior roles. Many people feel that to reach these positions, they have to "talk the talk."
- However, senior decision-makers, including the board, rely on reporting to inform and even guide them. If reporting is plagued by wordiness, woolliness, and evasive language, they cannot do so.
- Words have meaning; they can help or harm. How we express ourselves at work isn't merely an operational choice, but an ethical one.

NOTES

1 Orwell, "Politics," 110.
2 *Ibid.*, 105.
3 Grant, Adam. Think Again: *The Power of Knowing What You Don't Know.* London: Penguin Random House UK, 2021, 4. See also: Stephen E. Toulmin, *The Uses of Argument* (updated first edition) (New York: Cambridge University Press, 2003) and Jonathan Haber, *Critical Thinking* (Cambridge, MA: The MIT Press, 2020).
4 Bryan A. Garner, *Quack This Way: David Foster Wallace and Bryan A. Garner Talk Language and Writing* (Dallas: RosePen Books, 2013), 50.
5 Orwell, "Politics," 112.
6 Bill Birchard, *Writing for Impact: 8 Secrets from Science That Will Fire Up Your Readers' Brains* (Harper Collins Leadership, 2023), 22.
7 Todd Rogers and Jessica Lasky-Fink, *Writing for Busy Readers: Communicate More Effectively in the Real World* (London: Scribe 2023), 31.
8 *Ibid.*
9 Heffernan, *Wilful Blindness*, 47.

10 Orwell, "Politics," 113.

11 United States of America Department of the Treasury Office of the Comptroller of the Currency, *Report and Recommendation – Executive Summary Russ Anderson, Julian, McLinko* (occ.gov), 73.

12 Srinvisavan Ragothaman, Tyler Custis, and Melissa Christianson, "Fake Accounts Scandal at Wells Fargo: What are the Lessons?," *Journal of Forensic and Investigative Accounting*, 14:2 (July–December 2022), 315. http://web. nacva.com/JFIA/Issues/JFIA-2022-No2-11.pdf

13 Spicer, *Business Bullshit*, 13.

14 Orwell, "Politics," 116.

15 BP, *Deepwater Horizon* Accident Investigation Report, September 8, 2010, www. bp.com/content/dam/bp/business-sites/en/global/corporate/pdfs/sustainability/ issue-briefings/deepwater-horizon-accident-investigation-report-executive-sum-mary.pdf

16 Clarity Writing Experts, 2003.

17 See Simon Caulkin's explanation of the team's work in "Oh please, speak English," *The Observer*, August 24, 2003, www.theguardian.com/business/2003/ aug/24/theobserver.observerbusiness18

18 Berkshire Hathaway Inc., 2012 Annual Report, 3–4, www.berkshirehathaway. com/2012ar/2012ar.pdf

19 *The Big Short* (London: Penguin, 2011).

20 Antony Reuben, "HBOS risk control 'dumbed down'," *BBC News*, February 17, 2008, http://news.bbc.co.uk/1/hi/business/7892079.stm

21 Spicer, *Business Bullshit*, 21–26, 134–41.

22 "G. M. Internal Investigation Report," *New York Times*, June 5, 2014, www. nytimes.com/interactive/2014/06/05/business/06gm-report-doc.html, 253–54.

23 House of Commons Treasury Committee, *Fixing LIBOR: some preliminary find-ings. Second Report of Session 2012–13. Volume I: Report, together with formal minutes. Volume II: Oral evidence.* (London: The Stationery Office Limited, 2012), 65, publications.parliament.uk/pa/cm201213/cmselect/cmtreasy/481/481.pdf

24 *Ibid.*, 68.

25 House of Commons Treasury Committee, *Banking crisis: Regulation and super-vision. Fourteenth Report of Session 2008–09 Report, together with formal min-utes, oral and written evidence* (London: The Stationery Office Limited, 2009), 3, https://publications.parliament.uk/pa/cm200809/cmselect/cmtreasy/767/767.pdf

26 Financial Conduct Authority, Decision notice: Standard Chartered Bank, February 5, 2019, 1, www.fca.org.uk/publication/decision-notices/stand-ard-chartered-bank-2019.pdf

27 Enron, Annual Report 2000, https://picker.uchicago.edu/Enron/EnronAnnual-Report2000.pdf

28 Completely counterproductive, as demonstrated by Daniel M. Oppenheimer's wittily titled "Consequences of Erudite Vernacular Utilized Irrespective of Necessity: Problems with Using Long Words Needlessly," *Applied Cognitive Psychology* 20 (2006): 139–156, onlinelibrary.wiley.com/doi/10.1002/acp.1178 (Yes, the title is a joke – well, at least the part before the colon is.)

29 Birchard, 21.

30 Spicer, *Business Bullshit*, 11.

31 Spicer, *Business Bullshit*, 133.

32 "Garbage language: Why do corporations speak the way they do?," *New York: Vulture*, February 20, 2020, www.vulture.com/2020/02/spread-of-corporate-speak.html

33 Spicer, *Business Bullshit*, 45.

34 Office of Strategic Services (Washington, DC, 1944), 20, 23.

35 *Bad Language: The Use and Abuse of Official Language: First Report of Session 2009–10* (London: 2009), 3, https://publications.parliament.uk/pa/cm200910/cmselect/cmpubadm/17/1702.htm

36 Birchard, 29–30.

37 Orwell, "Politics," 116.

38 Young, "Garbage language."

Part 2

Words on the page

Chapter 4

Clarity

The practice

The previous three chapters have made a strong case for the importance of thinking hard about what you want to say, to whom, and why. Three chapters may seem a lot, but it's impossible to overstate the importance of understanding both yourself and your audience. Whether you call this topic critical thinking, analytical thinking, or something else, it's essential to writing. In fact, chief audit executives "indicate that analytical/critical thinking and communication skills are the most important" for their teams.[1] They won't be the only ones thinking this. Whichever sector, organization, or activity you work in, reporting requires rigorous thinking and clear communication.

But what is critical thinking? At its simplest, it is "[t]he objective analysis and evaluation of an issue in order to form a judgement" (Oxford Dictionaries), while the Stanford Encyclopedia of Philosophy defines it as "careful thinking directed to a goal." This may seem obvious, and many of us assume we do it every day at work. Yet assumption itself is a barrier to reasoning. Moreover, not all regions and cultures approach critical thinking in the same way. Depending on the country and its traditions, which affect educational practice, people may learn real-world applications through active learning, or a top-down, collective approach that prioritizes rote learning. Yet others feature curriculum content about critical thinking influenced by geopolitical views.[2]

This chapter will help you take the next step, putting words on paper (or screen). It will cover the basics – which I call the ABCs – as well as hindrances and aids to good, clear writing. These will include references to professional bodies and not-for-profit organizations offering resources and advice to their members and the public.

The previous chapters talked about two crucial influences on writing: cultural communication styles, and habit. Since these are learned and become entrenched over time, it is crucial for us to reacquaint ourselves with the principles of critical thinking and common barriers to reasoning. In the abstract to an insightful article entitled, "Teaching logic to auditing students: can training in logic reduce judgment errors?," Irvin T. Nelson, Richard L. Ratliff, Gordon Steinhoff and Graeme J. Mitchell state that,

DOI: 10.1201/9781003422365-6

Students in this study were taught formal and informal logic in an auditing course. They studied valid and invalid argument forms within the specific context of auditing services. These students, others without training, and a sample of professional auditors were then tested with a series of real-world auditing vignettes requiring critical reasoning and judgment. Students trained in logic outperformed students without such training. Furthermore, students trained in logic outperformed experienced auditors in their abilities to discern valid versus invalid argument forms. Conversely, experienced auditors outperformed trained students in their abilities to discern believable versus less believable argument premises.[3]

You may see yourself as experienced in your field, with a sense for faulty logic and gaps in evidence. However, all of us, no matter how intelligent or experienced, are human. We are just as liable to fall into comfortable patterns and use the reasoning that comes most easily to us. Often, we may be right – but are we at risk of missing crucial points that could lead us to different conclusions?

It's always worth refreshing our understanding of principal modes of (European) reasoning. The three examples below should seem familiar to you – but do you perhaps favor one when others would be more appropriate?

- Deductive reasoning uses verifiable facts to arrive at incontestable conclusions. "Everyone who works in this office is an auditor. Elvis works in this office. Therefore, Elvis is an auditor." So, if you know for a fact that all customer data is held in a specific database, you can state with certainty that if there is a flaw in customer data, you can direct your attention to that single database.
- Inductive reasoning strives for probable rather than certain results. However, those probable results are nevertheless based on sound, precise and verifiable statistics. "Ninety-five per cent of people who work in this office are auditors. Elvis works in this office. Therefore, Elvis is probably an auditor." Many of us in assurance functions reason in this way – absolute certainties are few, and we need to focus on the likeliest conclusion. However, be aware of this – and avoid making global statements such as "This system affects all customers" when you have evidence for only 90%.
- Abductive reasoning: "is a method of reasoning in which one chooses the hypothesis that would, if true, best explain the relevant evidence. Abductive reasoning starts from a set of accepted facts and infers their most likely, or best, explanations." (Wikipedia) "Elvis is most likely an auditor. He works in an office full of auditors, and enjoys making lists and organizing his office supplies."

I have personally seen many people in audit and risk functions rely on this last form of reasoning to speculate – based on extensive experience and common sense – about the root cause of a particular problem. "It's understaffing after the most recent cuts," they'll say, or "Staff haven't had the training yet." These are both plausible explanations – but in themselves aren't sufficient until you have sought both confirming and disconfirming evidence.

Without objective evidence, you may think you are being rational, without awareness of barriers to reasoning. Are you assuming something? Has precedent led to a bias or settled belief ("You always find these problems in HR.")? Have you mistaken correlation for causation? After all, staff may not have received training when they should, and you found problems shortly afterwards. This doesn't necessarily mean the two are linked – maybe the problems have arisen from a systems error, or the training itself is flawed and staff are better off not receiving it!

Adam Grant, in *Think Again*, emphasizes the importance of being aware of the gaps and flaws in our own thinking.

> In psychology, there are at least two biases that drive this pattern. One is confirmation bias: seeing what we expect to see. The other is desirability bias: seeing what we want to see. These biases don't just prevent us from applying our intelligence. They can actually contort our intelligence into a weapon against the truth. We find reasons to preach our faith more deeply, prosecute our case more passionately, and ride the tidal wave of our political party. The tragedy is that we're usually unaware of the resulting flaws in our thinking.[4]

Just as we want to communicate to people problems in their policies, procedures, and systems so they can fix them, we need to understand our own mental glitches and shortcuts. To write clearly, we must think clearly. And part of this surely includes asking ourselves, "What if I'm wrong?"

As with thinking, writing requires time and discipline. This eventually becomes easier with practice, as does any activity. If you regularly drive a car, cook food from scratch, or craft items from wood, over time you become more proficient and confident, able to perform competently in less time. This will happen with writing, too – provided you practice daily. If you apply the techniques in this book, and particularly this chapter, to everything you write, you will quickly discover three benefits. First, you will gain technical fluency – that is, the practice of plain language will become second nature, coming quickly and more easily as time passes. This in turn means you will produce clear writing more confidently and efficiently. Finally, it will make it easier for you to compile a report, as you pull together the components you have already written well.

To produce good reports, as with all writing, requires a relentless focus on using the fewest, *best* words, no matter the medium. It sounds easy, but it's not, and the best writers are those who take the effort to make their work readable. Again – hard writing makes easy reading, as exemplified in the famous quote attributed to everyone from La Rochefoucauld to Churchill: "I'm sorry to have sent you such a long letter; I didn't have time to write a shorter one."

Conveying complex information is hard; conveying it clearly and concisely is even harder. It's not impossible, of course – consider writer and broadcaster Jim Al-Khalili, professor of theoretical physics and chair in the public engagement in science at the University of Surrey. He communicates some of the most difficult scientific topics – quantum physics, for instance – in clear, reader-friendly prose. Whether in a book or a television program, the language he uses is not complex or flowery, because he knows readers will be focusing all their attention on the concepts he is explaining. He therefore chooses simple, elegant, transparent language that allows the reader to concentrate fully on the content.

This takes us back to the "fear" element discussed earlier; many people think that if they use simple language, it undermines the complexity or gravity of what they're saying. But the opposite is true: the clearer and simpler the language, the more easily your readers or listeners will understand your message. The aim is to deliver a message using the fewest, best words, so that the broadest relevant audience will understand it quickly and easily.

How do we achieve clear, simple language, especially when conveying difficult or complex messages? To make writing effective and efficient, we must focus on the essentials.

THE ABCs

These ABCs aren't the alphabet we all learn in our earliest school years, wherever we are. In this instance, they are the three crucial attributes of plain language in English: being active, brief, and concrete. I've referred to these as the ABCs for several years now.[5] It speaks to people's formative experiences of language learning while providing a handy three-letter acronym (beloved of most organizations). Most important, it helps all writers in English, whether native or non-native speakers, to avoid pitfalls and enhance their messages.

Active

By "active," we simply mean preferring active voice to passive voice in English. Other languages have greater linguistic and cultural tolerance of the passive voice; however, in English, overusing it can harm writing.

But what do we mean by "active" and "passive"? You can easily find explanations of varying degrees of accuracy and clarity, but it's best to focus on sentence structure. To put it simply, an active sentence in English starts with the doer – the person or thing performing an action. The action follows, and then the done-to (the person or thing on the receiving end of the action).[6]

Example: The manager reviewed the report.
Analysis: The manager (doer) reviewed (action) the report (done-to).

In a passive sentence, you simply switch the order around and move the done-to to the front of the sentence.

Example: The report was reviewed by the manager.

One important point is that we are talking about voice, not tense, here. Many people wrongly refer to the "active tense" and the "passive tense," which in grammatical terms makes about as much sense as referring to "sky biscuits" or "cat feathers." Tense is to do with *when* something happens – hence past, present, future, and many other tenses.

This means you can have active and passive voices in any tense. The example above was in past tense. Putting the same message into present and future tenses gives you the following:

The manager reviews the report. (present tense, active voice)
The report is reviewed by the manager. (present tense, passive voice)
The manager will review the report. (future tense, active voice)
The report will be reviewed by the manager. (future tense, passive voice)

Let's go back to the "The report was reviewed by the manager." In many cases, you'll see (or write) passive sentences that completely omit the doer. "The report was reviewed." This is perfectly fine, grammatically. However, if you get into the habit of writing in the passive voice, you will inevitably produce sentences without doers – and therefore entire documents with few clear authors or causes of the actions you describe. This will quickly become untenable.

Why?

It will become untenable for two reasons: first, it's harder for the reader to untangle what's happening. Second, if your writing obscures important information such as the doer, or cause, or origin, it may seem evasive. In English-speaking countries, for example, a favored expression of politicians or organizations hit by scandal – "Mistakes were made" – has become a joke. However, those hit by scandal haven't learned to be more transparent; they've compounded their folly by adding "and lessons will be learned." No one says who made the mistakes or who will learn the lessons. As a result, such phrases have become a hallmark of irresponsible, untrustworthy people and organizations – and note, these phrases rely on the passive voice. This is doubtless why George Orwell advised, "Never use the passive where you can use the active."[7]

Now, Orwell's advice is a bit extreme – "never" statements usually are. Rupert Morris, who provided the useful Clarity Index shown in Chapter 3, once told me that 80–85% of writing (in English) should be in the active voice. From decades of working as a professional writer and editor, and reading thousands of reports, I agree. Once your use of active voice drops below that threshold, you risk confusing or alienating your readers.

This means that up to 20% of your writing can be in the passive voice, and there can be many excellent reasons to use it. However, "Because everyone else uses it," "Because I don't actually know who is doing this," and "Because I'm in the habit of it" are not among them. It doesn't mean these reactions aren't common – they are. There's also the fact that many audit and risk professionals shy away from using the active voice because they equate it with blaming individuals. As Marisa Melliou, Group Audit Director, OPAP s.a., rightly says, "Putting the blame on a specific person does not usually help to communicate our findings and collaborate with business to find proper solutions."

She's dead right – yet how often would you indicate a specific person, by name? I have now read well over 3,500 audit, risk, compliance, and cybersecurity reports. Few mention individuals, and then only by role: the chief financial officer, for instance, or the division manager. These people are usually highly paid to manage their respective areas, and while we shouldn't personalize observations, there is nothing wrong with saying that the person in charge has not fulfilled his or her role.

Most reports will refer to a team, an area, a department, or a division, so any observations can state that "the team" or "senior managers," not named individuals. After all, readers know that the report is about operations in a particular area, or a certain team, so why would the writer suddenly become coy about it?

People are often too ready to accept passive voice because they say they don't want to upset anyone and derail relations with the client or business. This is understandable – as the previous chapters have made clear, conflict is something most of us avoid, and entire cultures develop ways to manage it.

It can go too far, though. I recently saw someone who advises internal audit teams saying that active voice can sound like blame, or be too harsh. This seems to me at best overly cautious, doing anything to avoid disagreement. At worst, it encourages people to think that avoiding difficult conversations is acceptable.

As mentioned in Chapter 3, the Global Internal Audit Standards, published in January 2024, lead with Standard 1.1: Honesty and Professional Courage. In an article on how the active voice promotes action, Richard Chambers and I pointed out that,

> We're not saying this is easy — simply essential. Internal auditors must be courageous when serving their organizations. Courage isn't just limited to "pushing against closed doors," or "sailing toward the storms."

We must also be courageous to tell it like it is when our work is done. Writing in the active voice often requires courage. But more importantly, it is often the call to action that our clients need to hear.[8]

Taking the easy option of using passive voice is understandable, but risky and potentially unethical. Overusing the passive is the perfect way to obscure any gaps or assumptions in your own fieldwork or research, as well as to hinder the organization in actually understanding and addressing problems. This is exactly why the passive voice is the go-to structure for anyone who wants to dodge responsibility.

It's hard to break the corporate habit of relying on passive. Try turning it into a game. One of my favorite ways of spotting a passive sentence is to add the words "by zombies" to the end of it. "The report was reviewed by zombies."[9] Have a go yourself!

When would you sensibly use the passive? You could use it if the doer isn't important. If, for example, all that matters is that the report be reviewed, then say, "The report will be reviewed." It could be clear from the context. If you say, "The manager reviews all reports. This one will be reviewed next week," then we know the manager will be doing the reviewing.

There could be legal reasons for using the passive – hence "Mistakes were made," where a company's legal counsel has probably strictly advised against saying, "We made mistakes." Then again, it could be that the writer doesn't know exactly who did something: "The firewall was breached at 10:37:26 on 25 May 2020." If the writer is updating senior managers about an investigation, it may be that this is all they can say at the time – they're still investigating the source of the breach (the doer).

So, there are several good reasons to use the passive voice. Habit, incomplete research, or fear of stating facts aren't among them.

There are even better reasons to use the active voice. If you are writing reports, you will be conveying information you have gathered yourself, or which other people have sent to you. In both cases, understanding the difference between active and passive voice can help you do your work better. Let's look at two examples, both taken from real (but anonymous) cases.

The first is from an IT report. In this report, the writers stated that "The IT Services Team does not adhere to Control A. Controls B and C are not adhered to, either."

This seemed clear on first reading: it was the IT Services Team that did not adhere to Controls B and C. Reading further, however, I became confused. I had thought one team alone was responsible for not complying with three different controls; the rest of the report suggested otherwise.

When I questioned the report-writers, they astonished me by saying, "Oh, a second team – not previously mentioned – is responsible for Control B. And to adhere to Control B, that team relies on a third team performing Control C. And no, we hadn't mentioned that third team before, either."

Why not? Well, according to the report-writers, because it was immaterial to anything except the recommendation for fixing the problem.

But it's not immaterial. Whether you're in audit, risk, compliance, IT security, or project management, you need to know where a problem arises and where it exists before you can solve it. If you phrase things in the passive, without stating the doer, you deprive the reader – and possibly yourself – of a clear view of what exactly is going wrong.

In this case, as I explained to the report-writers, you cannot recommend ways to fix a problem unless you know what it is. And part of knowing what it is consists of knowing exactly who is involved in what activities. So if, as I had originally thought, the IT Services Team was responsible for all three controls – and was not adhering to any of them – I'd assume the problem was within the team: recruitment, training, supervision. If an entire team is getting that much wrong, that's where the problem lies.

However, if – as we discovered in this example – you have three distinct teams who are all getting the same category of controls wrong, the problem may lie elsewhere. Is governance – roles and responsibilities, policies and processes, guidance and updates – clear? Are the teams relying on information from regulators in different jurisdictions, which can confuse matters? Are they using different systems, none of which interacts with each other?

Another example – one I've seen many times – is overusing the passive in a procedures document. Imagine, for instance, someone has just joined a customer services team and is studying the procedures manual. The manual may well include instructions such as these:

- When a customer communication is **received**, it is **scanned** into the workflow system.
- The communication is then **assigned** a case number and case handler.
- Requests are **sent** for briefings.
- Once the briefings are **completed** and **returned**, they are **scanned** into the workflow system.
- An investigative report is then **drafted**.
- Once the report has been **accepted** as being factually correct, a response to the customer is then **drafted**.
- The response is **reviewed**, **amended** as necessary, **signed**, and **sent**.
- The file is then **updated** and, if no further action is **required**, the case is **closed** on the workflow system.

I count 17 actions (in bold font) among these eight bullet points. Although this text isn't verbatim, I've based it on customer complaint-handling processes I've seen in different organizations. This kind of writing isn't unusual. So, what's the problem?

The problem is two-fold: first, for the poor new staff member joining the customer services team. How is this person supposed to know who does what? At no point is there a team name or job title to say exactly which actions belong to whom. In this situation, people will make mistakes – I've seen it happen. If they don't make mistakes, it's because they constantly check and double-check with colleagues and managers who exactly is responsible for what when. And that is a huge waste of time and resources – it's operationally inefficient.

The second problem is for people in assurance functions – auditors, risk, and compliance professionals – who need to assess such procedures as part of their reviews. How can anyone assess the procedure I've described above as adequate (designed to mitigate risk)? There simply isn't enough information.

The eight bullet points and 17 actions could, for example, be the responsibility of one member of staff alone. That's dangerous, though – leaving aside what happens if that person falls ill, there's a high risk of fraud or error from not segregating tasks and duties. That one person can make complaints disappear, or instruct compensation payments into accounts other than the customers'. It's a bad process and a bad control.

At the other extreme, the eight bullet points and 17 actions could be split among 12 people in six different teams. That's so complicated as to be useless – confusion is inevitable, as is duplication of effort. Again, a bad process and a bad control.

It could, however, be that it's a well-designed process, sitting within one or two teams and just enough people to promote checks and balances. In that case, a discussion with the manager should reveal this information. If so, suggest that he or she updates the process manual to make matters clear to everyone. However, if the manager can't or won't explain how the process works, take note – you may have some more investigating to do in that particular area! And all because the overuse of the passive voice alerted you to a potential problem.

It's easy to look at voice purely in negative terms, focusing solely on avoiding the risks posed by the passive voice. A more positive view is that using the active voice engages and draws in the reader. According to Louise McKay, UK Risk & Compliance Director at Royal London Group in Edinburgh, "Always use the active voice in preference to the passive – it makes a huge difference to audience engagement."

Using the active or the passive voice alone can make the difference between clarity and confusion, success and failure. This is an important point, and one we will develop further in Chapter 7.

Brief

Almost everyone has heard the advice to be brief when writing. Many people have approached me during training breaks, or at conferences, and barked a famous quote at me to show they agree with this point. Thomas Jefferson supposedly said, "The most valuable of all talents is that of never using two words when one will do." I've also heard these words attributed to Churchill and others. A more modern quote – and more easily attributable – comes from novelist Colson Whitehead, who advises writers, "Never use three words when one will do. Be concise. Don't fall in love with the gentle trilling of your mellifluous sentences. Learn how to 'kill your darlings,' as they say."[10]

But which darlings do you kill? Which words to cut? And why should you even try to do so?

You should be brief because it's easier to read. If you doubt me, consider the last report you struggled to read at work, and look at how long the sentences are. Most of them will probably exceed 15–20 words, a sensible guide advised by plain-language expert Martin Cutts.[11] This rule of thumb echoes that of Rupert Morris, who recommended no more than 20% passive voice – giving us another "rule of 20." Another way to phrase it is to encourage "20/20 vision" in your writing.

Using the active voice alone will help keep your writing brief, as active sentences are usually shorter than the passive versions. Consider the examples of the manager reviewing reports. The only way to make the passive sentences shorter would be to omit the doer – and that, as we said, creates another set of problems. So, when you "kill your darlings," start by targeting unnecessary passives.

If you don't keep it brief, through active voice and word choice, you will lose your readers. People in most workplaces are tired, distracted, and under pressure. A sentence that demands unbroken attention across two, three, even four lines is a burden. It's common for people to say that by the time they reached the end of a sentence, they'd forgotten how it started.

This isn't to say long sentences can't be readable; they can. However, it requires time, attention, and discipline from the writer to produce an impeccably constructed and punctuated sentence. In most cases, and especially when faced with a deadline, it's simply easier to cut a long sentence into two or three. (As stated earlier, if the writer isn't in a rush, he or she will probably be able to craft a shorter sentence to begin with.)

Even if you don't lose your reader's attention, you may lose his or her respect by using long words and sentences. As Daniel Oppenheimer of Princeton University proved, writers who show off are a turn-off.[12]

This brings us to why people write too much to begin with. In my experience, there are several reasons, all of which we need to address.

First, time pressure – it's easier to dash off a vague, rambling sentence than a clear, tight one where every word earns its place. It's like packing a suitcase by throwing everything into it and then trying to force it shut; the locks will break. Are you rushing and therefore cramming too much into your sentence-suitcases? Could you select fewer items, arrange them more tidily, and close the lid easily? Yes, but it's hard to do in a rush.

This is why famous writers have apparently apologized for not having enough time to write a short letter. It takes time, effort, and discipline most of us don't feel we have. But if the writer doesn't take the pains to write clearly, the reader will either give up trying to make sense of the writing or completely misunderstand it. The risks are clear.

A second reason is that the writer may not fully understand his or her own message. This is similar to people clearing their throats or using "Er," "um," "ah," "well," "you know," and other filler words when they have to speak unexpectedly. Writers asked to produce an urgent memo or brief are likely to start writing and think while they write. However, unless the writer then rigorously reviews and cuts, the resulting memo or brief is likelier to convey the writer's confused thought process than a crystal-clear message.

Finally, assumptions, beliefs, and habits often encourage people to write more. It may be that people think writing more will make them appear more intelligent, informed, educated, and professional. (The Oppenheimer research kills this myth.) It could be that they see others in the workplace producing long sentences and pages of verbiage, and think they must do so, too. It can be daunting and even scary, as we stated earlier, to swim against the corporate cultural current.

It often takes an outsider to see the corporate habits that inhibit clear writing. A belief that more is more, and that excess will impress, pervades many organizations. The investigative journalist Barbara Ehrenreich has worked undercover in companies to see how they work from the inside. For her book *Bait and Switch: The Futile Pursuit of the Corporate Dream*, she pretended to be a job-seeker. This allowed her to observe firsthand how the employment industry – including résumé advisers, personality-test advocates, and HR teams – works. Her insights show how people entering the workplace are encouraged to communicate from the start.

The advice she received on her résumé is telling:

> Break everything I claim to have done down into its smaller, constituent activities, so that, for example, I didn't just "plan" an event, I "met with the board to develop objectives" and went on through the various other phases of the job to "facilitate post-event evaluations." What can I say? It certainly fills up space.[13]

Speaking of filling up space, there is often an even simpler explanation for wordiness. In some countries, for example, most people are given minimum word counts for their assignments throughout their schooling. In many others, the word-count requirement disappears in favor of a page count by high school, at the latest.

What this means is that many people have had to continue to meet an often arbitrary word count six, eight, ten, or more years after their overseas colleagues. In this case, they will have acquired the habit of using words not to add specific meaning or nuance, but to hit a certain number of words. This habit will continue into the workplace. Instead of writing "now," people conditioned to write more for the sake of it will have written, "at this point in time." "We will do this" easily becomes "We will do this in the future going forward on an ongoing basis." Does any of this sound familiar?

The result will be pages of black marks, there simply to fill the pages with black marks. If your school, college, university, or professional body required you to hit a word count long after your age reached double digits, you will have to unlearn that habit. This conditioning takes years to overcome. It's worth unlearning, though, as your readers will benefit.

Another contributor to wordiness takes us back to culture. Many non-native speakers of English have two reasons to write more words than necessary. First, they may come from a culture where longer sentences are linguistically and culturally acceptable. When working with German clients, for example, I adapt Martin Cutts' 15–20-word guideline for English sentences. This is especially important because of German compound words, where a single word made up of multiple elements can create a wonderfully precise and nuanced term. My personal favorite is *Rechtsschutzversicherungsgesellschaften*, which all by itself conveys "insurance companies providing legal protection." You can see, though, how even a short sentence including two or three compound words could ambush the reader.

Other non-native speakers may tend to use longer words and sentences almost defensively. All too often I've heard people working in the UK, for example, say that native English-speakers they work with will automatically assume non-native speakers have a poor vocabulary or don't understand grammar. The result, though, is defensive writing, using the longest words and most complex sentences to say, "Look – I do know the language, and better than you!" It's a reaction I sympathize with. Having learned several languages myself, I've always gone through a phase of using literary or archaic (OK, pretentious) terms, simply to prove to native speakers that I could. I eventually overcame this tendency when native speakers asked me to proofread their writing – this showed I no longer had anything to prove.

The journalist Nesrine Malik writes beautifully of how her experience of learning, then mastering, English brought with it self-consciousness, prompted by (sometimes) well-intentioned criticism from others. Now, she says,

> I don't have time for that kind of preciousness about language any more. Having spent so many years trying to "improve" my English, I realised that the more I tried to follow norms, be they related to accent, pronunciation or even inflection and tone, the more hesitant and overly formal my English became. The English I ended up speaking is (as all languages are) dynamic and porous to other influences, and all the more expressive for it.[14]

Whichever language/s we speak, we all need to communicate. Using active, brief language will help you do so, without stripping your writing of its expressiveness. Earlier in the chapter, we mentioned certain metrics – a maximum of 20 words for a sentence in English, and a maximum of 20% passive voice across a document. You don't have to count these painstakingly yourself – software will do it for you. One of the most commonly available is the Microsoft Word readability statistics tool. You do need to configure this ahead of time in Word, so look online for guidance appropriate to whichever version of Word you have.

There are two important points: first, you must select "Grammar & Style." (Sometimes this is called "Grammar & Refinements," which is unnecessarily refined.) Second, you must review the entire document using both Spell Check and Grammar Check. Yes, it's tedious; however, when you are on version 27, it's after ten at night, and you've become word-blind, it will spot mistakes you made unknowingly. I've been writing professionally for over 30 years and still, when tired or distracted, will type "the" twice. I know I do it; I know I shouldn't; yet I do. It's called being human.

The wondrous thing is that once you have done all this, a text box will appear showing various statistics. The two you should focus on are under "Words per Sentence" and "Passive Sentences." If you can get both numbers below 20, your text is likelier to be reader-friendly. If not, go through that document, cut long sentences, and transform passive to active. You should then see the numbers going down, which can be encouraging. Remember, you're too close to the text to judge it objectively – let the software take some of the strain.

Readability Statistics   ?   X

Counts	
Words	1333
Characters	7785
Paragraphs	200
Sentences	51

Averages	
Sentences per Paragraph	1.6
Words per Sentence	14.6
Characters per Word	5.4

Readability	
Passive Sentences	7%
Flesch Reading Ease	38.7
Flesch-Kincaid Grade Level	11.4

OK

One word of warning, though. At the time of writing, newer versions of Windows feature Editor, which assesses and scores your writing. Beware the overall grade it gives as a percentage – my experience is that this "Editor Score" currently returns a lot of false positives. Check instead the readability statistics, which Editor should reveal, for your word-per-sentence average and percentage of passive.

Other software is available, Hemingway Editor and Grammarly being two of the best known at the time of writing. Both have the aim of alerting writers to common pitfalls so they can correct the text themselves. After all, the writer is usually best placed to respond to a comment such as "could replace with a simpler word." Speaking of simpler words...

Concrete

We've discussed the importance of using the active voice 80–85% of the time and keeping sentences below 20 words (in English). Toward the end of this chapter, I'll share tools and techniques you can use to achieve these goals.

The last element of the ABCs, though, is as important. It's being concrete, by avoiding vague, abstract language. One way in which writers can do this is by being alert to nominalizations, which Bryan A. Garner amusingly calls "zombie nouns."[15] Again – zombies to the rescue! What are they, though, and why are they a problem?

A nominalization is when you take a verb and make it into a noun. For example, "to analyze" is a verb; "analysis" is the nominalization (the noun form of the verb). To replace "to analyze" with a nominalization, though, requires you to add more words. You will have to say, "to perform an analysis" or "to undertake an analysis" to get the equivalent of "to analyze."

Garner's term "zombie noun" works well, as it conveys a lively image. Instead of a sleek, lively verb that conveys meaning and prompts action, we see instead a zombie: a sluggish, shuffling, stumbling brain-devourer. Martin Cutts also has a vivid term for nominalizations: "smothered verbs."[16] As you can see from the right-hand column below, the verbs certainly are smothered.

Using a verb	Using a zombie noun
analyze	perform or undertake an analysis of
implement	perform or undertake an implementation of
reconcile	perform or undertake a reconciliation of
compare	perform, undertake, or make a comparison of
approve	provide or grant approval of/to
buy	make a purchase of

You see in each case how the version on the right requires more words to produce the same meaning as on the left. Why would you do that? You'd be wasting words and trying the reader's patience.

You won't be aware you're doing this, much in the same way you aren't aware of overusing the passive voice. It will probably be a habit picked up in the workplace. As with the passive voice, it can create the same vagueness and potential confusion. A passive sentence without a doer is unclear about the source or cause of an action. In a process description, we saw how the passive obscured how many people or teams were involved.

The same can happen with zombie nouns. Some years ago, I saw a local government report on public housing. It included this sentence: "Retention and completion of all relevant sections of sign up checklists needs to be improved to evidence that arrears prevention is discussed with tenants from the outset." The sentence is over 20 words, has two instances of the passive voice ("needs to be improved" and "is discussed"), and, unsurprisingly, features zombie nouns: retention, completion, prevention.[17] It is unclear from the sentence *who* is meant to perform these actions. We can assume that someone or a specific team will discuss how tenants can avoid arrears; complete forms about the discussion; and retain them. We can also infer that someone needs to do a better job of the second two tasks: completing and retaining the checklists. *Who* is going to improve them isn't clear, though. Is it the same person or team who isn't doing it well to begin with? People higher up? A different part of the organization?

When writing is unclear – when it doesn't follow the ABCs – it can be hard to read, and even harder to critique. Many people complain about review processes in the workplace, and it's true they're often as ineffective as they are inefficient. Chapter 9 will cover the process, the pitfalls, and how to make reviewing a useful and positive experience for everyone.

However, think of the poor reviewers. Looking at the sentence above, we can see how much time and effort it would take for a reviewer to unpick the vagueness, gaps, and assumptions in it. No wonder few people take such trouble for every sentence in a 10- or 20-page draft, let alone a 400-page board pack. When reports convey important messages for the most senior decision-makers, or to the public, the consequences can be severe.

Chapter 3 featured an excerpt from one such document, the official report on BP's Deepwater Horizon accident investigation. It featured an uncharacteristically wordy and evasive passage: "Wherever appropriate, the report indicates the source or nature of the information on which analysis has been based or conclusions have been reached. Where such references would be overly repetitive or might otherwise confuse the presentation, evidentiary references have been omitted."[18] Let's analyze this passage using the ABCs. Is it as active, brief, and concrete as it could be?

Active: 100% of the sentences in this excerpt (both of them) feature the passive voice. Passive voice should feature in no more than 20% of the writing.

Brief: The average word count for both sentences is 20.5. This is only just above the 15–20-word guideline. It is also repetitive – "references" feature twice in a single sentence, ironically one criticizing repetition "would be overly repetitive."

Concrete: The excerpt features zombie nouns in passive constructions – "analysis has been based," "conclusions have been reached."

I suggested a different version in the last chapter: "This report gives the source and nature of information wherever possible. Where we feel evidence is repetitive or potentially confusing, we have put it in the appendix." Zero passive voice; 13.5 words per sentence.[19] And understood instantly.

More importantly, the active voice and especially the word "we" create a relationship with the reader. Rather than pronouncing from a lofty position on what the reader can or cannot understand, the rewrite tries to say, "This is what we've done and why. We hope it's useful and easy to read." As Louise McKay said, it engages the audience.

Even resources meant to help people communicate more clearly may not help. The book *Office English*, from 1961, aims to accompany new employees in an office environment as they discover such novelties as tape recorders and telegrams. Much of the advice is practical and sound. The preface, however, says,

> The aim of this book is to produce facility and accuracy in expression, with particular attention to the use of English in commerce. Exercises for increasing the range of vocabulary, extending the scope of sentence-patterns and building substantial and well-arranged paragraphs are given. Attention is paid to spelling and to those rules of grammar having direct bearing in clarifying correspondence.[20]

How many instances of the passive voice can you find in this quote, and how many zombie nouns? You may think that the date of the book excuses such a wordy, pompous style. However, as you'll see in the resources list at the end of this book, several authorities predate *Office English* by several decades. All warn against reader-repellent writing.

As Rogers and Lasky-Fink advise,

> You will rarely go wrong with a modified version of "less is more": Aim for the least amount of complexity that will allow you to engage your intended reader. You need to pay attention to context, but remember that more readable writing is fundamentally more effective writing.[21]

Many writers, especially those lacking confidence, may be tempted to use large language models (LLMs) at this point. After all, if software such as ChatGPT can draft text for you quickly and easily, why master the discipline of plain language?

For those in assurance functions, it seems natural to use the available technology to speed up the process. As Alee Marschke says, "Brainstorming, critical thinking, and writing are key requirements for an internal audit. These skills blend smoothly with the capabilities of AI Large Language Models (LLMs), making internal auditing a viable space to incorporate AI tools into everyday work."[22]

Many people feel nervous about new technology, and LLMs are no different. Software such as ChatGPT evolves constantly, using prompts and text fed to it, to improve its performance. As it improves in response to human input, though, its outputs may become – to some people – worryingly human-like. As Marschke states,

> For some, a tool based on human knowledge may lead to more natural, intuitive interactions that allow for a more efficient exchange of information. For others, the blurred line between AI and human interaction may increase fears of the inability to distinguish between the two or add to worries regarding the sentience of AI.[23]

Lorelei Lingard writes about writing, specifically in medical education. In an illuminating article, she tested ChatGPT to see its uses and pitfalls, and experienced first hand what she perceived as its human-like aspects: "I know it isn't sentient and doesn't have motivations or emotions, but I can't help but think in some of our exchanges that it was being sullen, intractable, even deliberately insincere."[24]

Yet people with backgrounds in accountancy or IT, for instance, may feel that writing skills are not their forte – numbers and coding are. However, everyone can improve their writing skills, and we cannot outsource or abdicate responsibility for clear communication. As with any technology – typewriters, word processors, smartphones, or LLMs – human beings must take the lead in prompting and checking any text produced. After all, even if you use a traditional typewriter, you must think about what you are typing and check for manual errors afterward.

If you choose to use LLMs, think about what you include in prompts. First, plain language – and the critical thinking that should precede and accompany it – is crucial to creating prompts likely to produce relevant text. According to Alexander Rühle, CEO and Co-Founder of Zapliance,

> generic prompts equal generic output; not usable for assurance but for brainstorming purposes. Specific prompts are necessary. To get to working prompts, shifting mindsets and retraining are necessary. Writing regularly is a process to derive a logically complete text. When prompting, you need to know beforehand the characteristics of the output.

Equally important is the content of your prompts. As IIA Global has stated, "internal auditors should not include any proprietary information about their organization in their prompts. Remember: ChatGPT stores everything that is typed into it."[25]

Further consequences of LLMs' use of input have appeared. Following the *New York Times'* lawsuit against OpenAI and Microsoft for copyright infringement in December 2023, eight US newspapers followed suit in April 2024.[26] Entering data into LLMs may generate useful text speedily – but it can lead to legal and reputational harm, as can using LLMs without attribution when producing text.

A third point is that while LLMs may produce useful text, they may also produce nonsense. As Lorelei Lingard states,

> ChatGPT is a text generator, not a brain: it is putting together words that are likely to be found together around the topic you've asked about. That doesn't mean these words 'belong' together or that they are 'true'. In fact, ChatGPT seems to enjoy making sh*t up. You absolutely cannot trust the references it gives you.[27]

OTHER HINDRANCES TO CLEAR WRITING

Personal hang-ups

We've been reading words since the start of this book, and I've shown how to put the ABCs – words – into practice. So, what else do we need to say about them here?

Everything.

Words are emotive, and not only ones that indicate anger, contempt, criticism, and the like. How we use words, pronounce them, and spell them says a great deal about us. Most people can immediately think of national or regional assumptions. Many cultures associate accents, dialects, and slang negatively with the speaker's education and even intelligence. Understanding that a country or language can have different registers makes the difference between engaging and alienating people. Words can hinder as well as help, and how we react to other people's use of them often reflects more about our own beliefs and insecurities.

Some people like to pretend that language – and words – are irrelevant. I can immediately think of two such instances from when I worked in a global bank. In each case, the colleague – a native English-speaker – asked me to review a draft document for clarity. One document was an email, the other a set of slides. Both documents sought to convey important information about projects.

These colleagues never knew each other and didn't work in the same area. What they had in common was the way they asked me to review what they had written. Both said exactly the same thing: "Can you have a look at this and tell me if it makes sense? But don't bother about the words or language or anything – just the message."

You can see the lack of logic in this request. My response was to ask them how they planned to convey the message, if not through language. Cave paintings? Interpretive dance? Mime? What they'd said made no sense, and yet it speaks to some underlying truths about people and wording.

When I reflect on it, they asked the same question, but came from different perspectives. The first person who asked me came from a scientific background and proudly asserted that she "wasn't good at English" (her native language). She genuinely didn't have a problem with errors in spelling, punctuation, grammar, or usage "unless they affect the meaning." But, because she was indifferent to the accuracy of the written word, she wasn't alert to potential problems.

The second person who asked me this was aware of the importance of language, but defensive. She spoke with a strong regional accent that some colleagues mocked. As a result, she often felt judged for speaking the way she did, as so many people couldn't overcome their prejudices to listen to what she said, which was almost always sound. She was an intelligent and well-informed colleague, but, because of her experiences, she anticipated negative responses to anything she said or wrote.

You may know people like this in your organization. If they're senior, it can be difficult to overcome their habits and assumptions about communication. However, you have an opportunity to discuss with them how important words are, and how to use them judiciously.

What may encourage you is to know that at the most senior levels of organizations, most people appreciate clear language. Sometimes it comes as a relief when a new director sets the standard for plain language. One such is David Dart, Head of Allied Command Operations Internal Audit at NATO. "I brought with me my own style of formatting and writing audit reports," he says, "and I have adopted this across my team without any complaint." In fact, his team welcomed his approach, especially as it favors plain language: "I particularly dislike long sentences and complexity, and, although a degree of formality is important, it should not cloud the message." It is, after all, possible to be professional and plain – in the best possible way. But all too often, people confuse formality with pompousness, aided by corporatespeak.

Corporatespeak (yes, again)

This brings us back to the language discussed in Chapter 3. Molly Young, however, prefers the term "garbage language," which is nothing if not emotive! "I like Anna Wiener's term for this kind of talk: *garbage language*," she writes.

> It's more descriptive than *corporatespeak* or *buzzwords* or *jargon*. *Corporatespeak* is dated; *buzzword* is autological, since it is arguably an example of what it describes; and *jargon* conflates stupid usages with specialist languages that are actually purposeful, like those of law or science or medicine. Wiener's *garbage language* works because garbage is what we produce mindlessly in the course of our days and because it smells horrible and looks ugly and we don't think about it except when we're saying that it's bad, as I am right now.[28]

Chapter 3 concurred with Young's distaste at the lazy, vague, often meaningless terms people use as linguistic shortcuts or camouflage. If you don't know what you want to say, or don't want to say it clearly, these terms will help you in the short term. In the medium to long term, though, as we saw, they prevent organizations from operating effectively and transparently. I am happy enough with "corporatespeak," a term I've used throughout, but am open to other suggestions.

After all, why not BAU-speak? George Orwell invented the term "Newspeak" for his novel *1984*, which showed how well totalitarian regimes appreciate the power of language to manipulate and deceive. For those interested in social sciences, Alan Ross created the term "U-speak" for the language used by the English aristocracy (the upper classes, hence U).[29] Why not BAU-speak, so that, in keeping with politics and sociology, business too has its own term?

Most people understand the three-letter acronym[30] BAU to mean business as usual, which is exactly my point. Whenever middle managers use the latest linguistic fad, they aren't being creative or innovative – quite the opposite. They are engaging in business as usual: the tired practice of trying to inflate everyday tasks, or obscure poor performance, through inaccurate, off-putting, often pretentious language.

Louise McKay has pinpointed how seductive and dangerous this practice is when she highlights "the tendency of all organizations to lapse into jargon and corporatespeak. People assume that there is a common understanding of that language. However, it's surprising sometimes to find how hazy certain definitions are." This goes back to my point about "crutch words" in Chapter 3 – terms people use automatically, reaching for them out of ease rather than deliberate choice. We all have them – becoming aware of them is the first step to reducing their use and power.

I could spend several pages dissecting some of my current least-favorite corporatespeak terms. Ones that come immediately to mind include "onboarding,"[31] "upskill," "value-added," and "paradigm shift." Since I remember people mocking this last term back in the 1990s, it can hardly be at the cutting edge of expression – hence BAU-speak. However, writers including Steven Poole and Don Watson have devoted entire books to the topic, while Martin Cutts' *Oxford Guide to Plain English* includes a table of "plain words," encouraging simple terms in favor of more archaic or pompous ones.[32] For internal auditors, the classic *Sawyer's Internal Auditing* devotes three chapters to reports and includes a list of plain words. "Write to communicate your findings and express your ideas," the authors rightly say, "not to impress someone with your knowledge."[33]

One category of corporate jargon many people struggle with is associated with business methodologies. Many of you will have heard of – or have qualifications in – Six Sigma, Lean, Agile, and other approaches to process improvement and project management. My own experience of this kind of thing dates back to PRINCE 2, which I found reasonably clear and sensible. Since then, however, an entire industry has grown up around creating hierarchies, training, and a specific language for different ways to do things. Often the approaches themselves come from environments where they make perfect sense. Translated into completely different settings, however, what thrives is the terminology: green and black belts, scrums, and buttered badgers (I made that one up).

It's not entirely clear to me – or to many researchers who have devoted their lives to studying this – whether these approaches produce better results. It can be hard to discern such things when the vocabulary alone distracts and often irritates. If you have to review an area or project that uses an "agile" methodology, you will probably have to refer to some of its terminology. Keep the reader in mind, though – parroting the latest management fad rarely conveys useful information.

The resources at the end of this book include the works cited above – Cutts, Poole, Watson, and others – all of which promote clear writing and meaningful reporting. These books will help you not only expand your vocabulary, but also build your confidence. You alone are responsible for conveying your message clearly, and the more options you have, the greater your scope. According to David Foster Wallace,

> In order for your sentences not to make the reader's eyes glaze over, you can't simply use the same core set of words, particularly important nouns and verbs, over and over and over again.... Having a good vocabulary ups the chances that we're going to be able to know the right word, even if that's the plainest word that will do and to achieve some kind of elegant variation, which I am kind of a fiend for.[34]

Not all of us can achieve elegant variation – but at least we can try to avoid making our readers' eyes glaze over.

This isn't to say that you should try to use unusual words, or seek to impress the reader with your verbal acumen. After all, even if you had a sports car capable of achieving the highest speeds, you wouldn't always drive that fast. It's about having the widest selection to choose from, so that every word does exactly what you and the reader need.

Remember – the English language is global. This means that it belongs to all of us. Many people – especially native speakers in the UK – have told me they don't feel confident using their own language. Why? Maybe they speak with a regional accent, or lapse into dialect, as the colleague I mentioned earlier did, and were mocked. Some say they weren't good at English at school, learning only later that they had dyslexia. Some believe that only people who study English at university can claim to "be good" at writing in English. (Guess what – I didn't study English at university.)

Let this book comfort and encourage you – English belongs to you and indeed to all of us. And using it consciously and in good faith enriches the language for everyone. Corporatespeak enriches nothing. So, if you want to use language more clearly and usefully, look to the resources.

Mechanical mistakes: Basic accuracy

Please don't be offended. You may well be the person every team needs, who points out that buildings can be evacuated, but people can't (unless under medical supervision). You may be the one whose teeth grind every time you see someone using "actionable" to mean "feasible" or, even better, "doable." You may think "uninterested" and "disinterested" are always interchangeable. (Maybe you don't care, because you don't see how it could possibly affect you.)

However, if you're not that person, you may well want to think about these points. Most people assume that they generally match their words to their meanings, and they usually do. It's extremely easy, though, to assume a word you often hear is accurate – or even a word. I've read dozens of reports that have had me wondering how I could have missed the invention of a particular word everyone has started using. Sometimes I have to look up whether the primary meaning has changed – "actionable" is a perfect example. (Answer: It hasn't. If your report proudly lists events you claim are "actionable," you should have a good lawyer on retainer.)

Rob Reinalda is an experienced writer and editor, and author of *Why Editors Drink*. One of the things he highlights is the complacency that leads people to write without checking.

> If only there were reference materials that explained the distinction. ... Looking stuff up used to require getting up, walking over to the dictionary, flipping through its pages (maybe puzzling over the whole alphabetical order thing), finding the word, and often cross-referencing it – consulting a different book, a thesaurus, to ensure you're using exactly the right term. Now, you simply type a word or phrase into a search engine, and an array of options will pop up. With a few clicks, you can see the words' definitions, etymology, synonyms, antonyms, Cinnabons, Auntie Em – yet people refuse to double-check for proper usage.[35]

Misused words aren't the only trap lying in wait for the unwary (or indifferent) writer. One category of words – contranyms, or Janus-words – can seriously mislead readers. These words have two definitions, each the opposite of the other. Several of them are common in reporting, which makes it even more important to be alert to them. For example, "sanction" can mean to approve or to forbid, because its fundamental meaning is to enact something by law. If you say that senior managers have sanctioned changing a process, do you mean that they have permitted or banned it?

Another example is "oversight," which usually means "supervision" in reports. However, it can also mean a lapse in attention, a mistake. You could therefore have a report that says, "This error occurred because of management oversight. It will be corrected because of management oversight." (Note that the second sentence also features a passive. Bad habits of writing tend to hang out together, like a gang of word-thugs, ready to bludgeon the poor reader into submission.)

If your report reviewer or line manager is the kind of person described in the first paragraph, consider what you may be unintentionally inflicting on them. There are few things more irritating than having to correct vocabulary in a report, not because of personal style, but because of basic accuracy. Words do have meanings, and we have ways of checking them. Bookmark – or buy – at least one reputable, standard English-language dictionary for everyday reference. The most trusted include Cambridge English, Chambers, Collins, Merriam-Webster, Oxford English, and Random House dictionaries. The websites for these works also offer thesauruses, which are helpful when you want to vary your wording, or search for the exact shade of meaning.

Mechanical mistakes: Grammar (including punctuation)

My original plan had been to devote a separate chapter to grammar. However, an early reviewer suggested including the topic in this chapter, given how many resources are available elsewhere. The lengthy resources list at the end of this book will prove the reviewer right.

Yet it's still worth covering *why* grammar regularly features in most books and training about report-writing. Most people should have covered the topic in elementary or primary school, yet many of them will claim to still be confused. This could be because the grammar teaching they received was poor or nonexistent. We can include in this category exceedingly rigid approaches, as these usually indicate a superficial grasp of the subject and thus an unhealthy obsession with a few "rules." (Yes, I will explain why I wrote "rules" shortly.)

This takes us back to the concepts of self-consciousness and fear when communicating in the workplace. Many people have experienced managers wielding red pens, or tracking changes, across draft reports. Sometimes the managers' comments will be accurate or useful. More often, the comments will betray confusion about the purpose of a review, as well as a deep-seated need to impose one's style on others.

We can't organize counseling for every manager who thinks "reviewing" means "rewriting." What we can do is recognize some important differences. First, I've heard many managers say they have to rewrite people's reports because "of all the glaring grammatical mistakes." When I look at the reports in question, there are never as many "glaring grammatical mistakes" as I was told. What I do see are numerous unnecessary stylistic changes, none of which changes the meaning. This is what people mean what they complain about managers and other reviewers "wordsmithing" reports.

Now, a "wordsmith" should be a fine thing, like a goldsmith or silversmith – someone who creates beautiful and useful things from precious materials. The term "wordsmithing" in the workplace, though, usually means people who endlessly carp and cavil over wording to no good end.

What is more telling is the number of times the "wordsmithing" overlooks basic errors in spelling, grammar, usage, and punctuation. In other words, the same managers who complain about their teams' poor grammar can't even see it.

Why does it matter? After all, many people will say that the only people who care about grammar and punctuation are the managers who claim to be experts in it, and then use that claim to browbeat their colleagues. No wonder it puts people off.

It matters because it can confuse or obscure meaning. Consider a sentence frequent in banking: "This incident has affected all of the customers accounts." Without an apostrophe, the reader has no idea if the incident has affected all of the accounts belonging to all customers ("the customers' accounts") or all the accounts belonging to a single, but extremely important customer ("the customer's accounts"). Similarly, if someone sends an email saying, "Let's go break the managers legs," readers wouldn't know if the sender proposed maiming only one manager ("the manager's legs") or several ("the managers' legs"). (I hope they would call the police, though.)

It's not just the oft-dreaded apostrophe that can muddle meaning. Misplaced modifiers are a common source of misinterpretation. (Quick aside: a modifier is a word or phrase that modifies another word or phrase. It should – in standard English – be placed as close as possible to the thing it modifies. Remember: "Modifiers are like teenagers: they fall in love with whatever they're next to."[36])

Spot the difference in meaning in the following four sentences, all featuring the modifier "only".

- **Only** we reviewed the IT security governance framework.
- We **only** reviewed the IT security governance framework.
- We reviewed **only** the IT security governance framework.
- We reviewed the **only** IT security governance framework.

The first sentence states that we are the only people to have reviewed the framework – no one else did. The second sentence means that the review was the only action we carried out – we only reviewed it; we didn't design it or manage it. The third sentence says that of all the things to review, the only one we did review was the framework. The fourth sentence makes clear that there is only one framework – no others.

If any of this seems confusing, remember to consult the resources list, which includes books and websites that go into more detail, give many examples, and even have interactive exercises and quizzes. However, you can also use a technique common to professional writers the world over: read your text out loud. This will help you spot errors that spelling- and grammar-checkers may overlook – "colleges" where you meant to write "colleagues," for instance. It will also alert you to anything confusing or ambiguous (such as a misplaced modifier). My tip is to read aloud and trust yourself. If something sounds fine, it probably is; you've likely created a simple, clear sentence that won't frustrate the reader. If you stumble over a word or sentence, though, it's a sign you need to change something in it.[37]

Everyone – including professional writers and editors – makes mistakes. It can be because we're tired, distracted, or simply never felt confident spelling "accommodation." What we need to do is be alert to the possibility, use any resources or help we can, and correct mistakes before they go out in a final version.

Making mistakes is not in itself morally reprehensible (although some people's reactions to a typo would make you think so). The problem is that at worst, they can confuse, and at best distract the reader. I recently read a book that, while extremely useful, entertaining, and well written, had so many basic errors in spelling, grammar, punctuation, and usage, that I couldn't focus. The copyediting process at the publisher had clearly failed, but it was the reader who suffered.

There can also be more serious consequences to grammatical mistakes. Cecelia Watson, in her brief yet brilliant book *Semicolon*, devotes two chapters to legal cases arising from the (mis)use of that poor, misunderstood punctuation mark.[38] There are also recent cases where plaintiffs relied on the interpretation of a punctuation mark to claim, sometimes successfully, millions of dollars.[39] These can be amusing to read – unlike the cases where a misplaced mark meant life or death.[40]

There are many people who believe grammar is *always* a life-or-death matter. They are also often wrong, as the linguist David Shariatmadari says. In his book *Don't Believe a Word*, he devotes a chapter to the eternal and tedious chorus of people who bemoan what they see as poor use of language. Time and again, as he and others have proved, their very complaints show their ignorance. "If they're so concerned about language," Shariatmadari writes, "you have to wonder: why haven't they bothered to get to know it a little better?"[41]

The next time someone tells you that you can't start a sentence with "but," or put a comma before "and," or split an infinitive, refer them to some experts on the subject. Martin Cutts, in his *Oxford Guide to Plain English*, has an entire chapter on language myths. Benjamin Dreyer also tackles some widespread but ill-informed beliefs in his book, *Dreyer's English: An Utterly Correct Guide to Clarity and Style*.

This comes down to understanding what grammar is. It is how we organize our language, and is made up of syntax (word order) and morphology (how words change in sentences). As language changes, so grammar changes. Jane Austen regularly used apostrophes in her possessive pronouns, whereas nowadays we never do. Instead of "rules," we can perhaps talk about "guidelines." Instead of talking about "proper" English (as people who get upset about split infinitives, without knowing what they are, tend to do), let's say "standard." Grammar snobs will always be with us, claiming there was once a golden age in which everyone spoke and wrote perfectly accurate English. As if.

English belongs to all of us. There is no excuse not to become better informed and more confident. Look at the resources at the back of this book, as well as any others people recommend. Use what works for you – if you find a particular author's tone intimidating, or a website hard to navigate, use something else. There are too many excellent works out there not to be able to find something, or many things, that are right for you.

Mechanical mistakes: Usage

If you think grammar is contentious, consider "usage," defined best by Jack Lynch as "how to use something properly."[42] You can avoid most mistakes in usage by relying on the resources I've mentioned throughout this chapter: dictionaries, thesauruses, and guides to clear writing and plain language.

You will always encounter people whose views on usage differ from yours. This may be down to national or regional differences.[43] When I worked in Edinburgh, many colleagues used the Scots term "outwith" instead of "outside." My view was that this was perfectly fine for communications among Scots. However, the bank was a global one, meaning that region- or nation-specific terms quickly confused colleagues from other locations. Therefore, I reasoned, any communications sent to people outside Scotland should use the standard English term "outside." Although some people resented having to adapt their vocabulary, they understood the purpose.

There are also people who – similar to the grammar snobs – categorize their preferred terms as the correct usage and everyone else's as poor usage or even wrong. Chapter 9 will discuss in detail how these and other factors can complicate the review process. For now, though, you should explore the recommended resources – especially dictionaries – and use them to have a discussion. Often people with fixed ideas about "proper English," including grammar and usage, are astonishingly unaware of the very things they profess to love and defend, as Shariatmadari made clear in the quote above.

One topical point of usage is inclusive language, which was mentioned in the previous chapter's activity. In French, many people are adapting their writing to include both female and male in gendered terms. This is because the traditional approach was to subsume the female into the male, even if the female was greater in number. For example, a document about customers would use the term "clients" (masculine plural) even if there were 99 female clients ("clientes") and only one male. Nowadays, you will see documents using "client·e·s." There are, of course, debates about this – explicit and deliberate changes to language rarely happen without an argument. However, it illustrates how language can be seen to exclude or include.

In English, we haven't had gendered nouns for some time. However, until recently, the default pronoun when speaking of a person was "he" – it was assumed that the male would include the female. Many women rightly objected to being subsumed in this way. But how to solve the problem? Common options include using "he or she," "s/he," or even alternating "he" and "she." So the French document about clients would, when speaking of a hypothetical singular client, say "He will expect good service," followed by, "She also considers value for money." This can be confusing, though.

Another option is to use "they": "When a client walks into a shop, they expect good service. They also consider value for money." It's common for people to claim that you simply cannot use a plural pronoun ("they") to refer to a singular. Yet for centuries, people have used "they" in this way.[44]

Consider your audience, refer to your organization's style guide (if it exists), and decide for yourself. Whoever reads your report, they are sure to appreciate the effort.

Finally, if you are ever tempted to use gendered language for occupations – don't. Even the 19th-century author Ambrose Bierce found it ridiculous: "Poetess. A foolish word, like 'authoress'."[45]

HELP IS AT HAND

Earlier, you saw how tools such as Microsoft Word readability statistics, Hemingway Editor, and Grammarly can help you improve reports. Wherever you can reliably use technology to alert you to clunky or confusing wording, do. It's easier and often more accurate than trying to do it yourself, especially when you are tired and too close to the document.

There are also other, less mechanical, resources. First, have you looked at your professional body's website? Many offer advice, links, and sample documents tailored to your everyday work. Keep in mind that resources are often available only to members, so you may need login credentials to access the organization's materials.

You may be surprised by how professional bodies address not just reporting standards, but also wording. The American Institute of Certified Public Accountants has a Center for Plain English Accounting site (www.aicpa.org/interestareas/centerforplainenglishaccounting.html), as well as workpaper templates available through its main site (https://future.aicpa.org/home). Meanwhile, Chartered Accountants Ireland has produced advice on reporting to third parties, including "examples of types of engagement wording or of opinions that are unacceptable to accountants providing reports."[46] Although produced for Irish accountants, this type of advice can be useful to internal and external auditors, as well as risk managers, across the globe. Much of it is sound common sense and awareness of language.

If you belong to a professional organization or body, you will doubtless have standards, including a code of ethics and guidance for carrying out your work. The Institute of Internal Auditors' Global Internal Audit Standards task the chief audit executive with promoting "accurate, objective, clear, concise, constructive, complete, and timely internal audit communications" throughout an engagement.[47] If the ABCs described above – being active, brief, and concrete – can help you meet this standard, how else can they improve your writing?

It's worth casting your resource net more widely. Maybe your professional body doesn't offer up-to-date advice, or guidance relevant to your organization. One person who thinks creatively about how to make reports both compliant and effective is Komitas Stepanyan. As Technology and Cybersecurity Director at the Central Bank of Armenia, Komitas is well placed to appreciate both the complex information his team needs to report and their readers' needs. The first point he makes takes us back to the ABCs, with the need to be brief:

> There are different types of reports and the report-writer needs to clearly understand "who is the audience?". The higher your position, the more reports you will have. If you have more than ten reports, and if every report is more than two pages, then mostly probably you will not be able to read all the reports. Statistics show that many top managers do not read reports more than 1.5 pages.

To achieve clear, compelling reporting, Komitas urges his team to think creatively and use their imaginations.

Where can you find different approaches to inspire your own thinking, and create new ways to improve your reporting? There are many sources of writing advice on the internet, along with training and coaching services. Depending on your country, language, culture, professional sector, and area of expertise, these can range from irrelevant to extremely useful.

Plain language is a common goal in many organizations, explicitly so for many federal or local government bodies. However, many people assume that plain-English or plain-language entities using "campaign," "foundation," or similar words in their names must be objective, not-for-profit, or other public-interest organizations. This is not always the case. However, two of my favorite resources are indeed not-for-profit organizations: Plain Language Association International, which is staffed by volunteers, and the US Government's Plain Language site, staffed by (paid) government employees.

I mentioned in this and previous chapters how personal communication and writing styles are. This means that anyone who wants to help others communicate well should be able to do so themselves. This may seem obvious, but many of you will have experienced well-intentioned managers who cannot write well advising – or dictating – how others should write. It's a common human failing – haven't we all heard the famous "Do as I say, not as I do"?

In trainers and consultants, however, there is an ethical as well as a logical problem. Going back to the link between language, logic, and morality, it is fundamentally dishonest to charge people or organizations for advice on a skill one does not have. There are report-writing trainers whose own writing is unreadable or includes sloppy mistakes, and communications consultants who seem oblivious or indifferent to corporatespeak. Similarly, someone who criticizes other people's usage and issues lists of "banned" or "required" terms probably doesn't understand the history of language. Language – any language – is too rich and complex for anyone to reduce it to a list of personal habits and beliefs, let alone present this list as universal truth.

If a writing consultant cannot *do* what they advise others to do – or worse, cannot see the problem with their inability – then they cannot claim to be competent. Similarly, someone who claims to be an expert in cross-cultural communication but speaks only one language (usually English) may not even be aware of the nuances and insights overlooked. This can introduce risk into communications, rather than reduce it.

If careless, vague, or clichéd language undermines credibility in reports, it does so tenfold when used by anyone who professes to be an expert in clear writing. So, if you are seeking professional expertise and advice in report-writing (and there are many excellent providers), check first how that organization or individual uses language. If you see "upskill," "onboarding," "value-added," or similar linguistic red flags, look for another provider. After all, why would you pay anyone money to advise you on the very thing they cannot or will not do?

HELP YOURSELF, HELP YOUR READERS

The advice and resources provided in this chapter are simple, but not easy. As we've said several times, it takes time, effort, and discipline to rigorously analyze our own thinking and carefully choose the most precise way to express it. Many of the reasons why people shun good practice are familiar to use from Chapter 3. Habit, tiredness, and fear all conspire to make people produce unreadable writing.

Being aware of these factors, and consciously working to overcome them, will produce better writing every time. If you apply the content of this chapter to everything you write, it will improve your skills and make every subsequent task easier.

Earlier we used the analogy of building a house: clear thinking and plain language are the foundations, without which any house will fall. Another house analogy applies to reviewing your own or someone else's work. If you want to clean your house, or do any other work in it, it's better to clear all surfaces of clutter first. Similarly, if you declutter your writing, you and other readers can more easily do the work of grasping the message and taking useful action.

It's hard work, and it can sometimes be even harder to get colleagues to accept clear writing. It requires resisting corporate habits and overcoming individual fears. However, if you or others fear that clear writing and plain language may come across as rude, try to reframe the problem. A clear message saves your reader time and puts him or her in a better position to address gaps and failures. It also makes the organization's governance stronger, more transparent. As John Chesshire says, he expects reports to be

> clear, concise, evidence-based, jargon-free and active, please. I have a small brain, so I need to be able to easily understand what is being said. I want to see the important things highlighted, and practical, pragmatic, relevant solutions recommended and agreed. The reports have to generate an enthusiasm for action and improvement, where necessary. I want people to read our reports and to keep turning the pages to the end.

One carefully chosen, precise word is better than masses of words shimmering coyly around the message. Your readers' time is limited, their attention spans overloaded. Keeping your words few and judicious increases the chance people will keep turning the pages to the end – and then act.

ACTIVITIES

- Look at a recent piece of analysis you wrote - it can be an investigation, review, or audit. Tracing your work back to its beginning, what types of reasoning did you use - deductive, inductive, or abductive? Why?
- During your next meeting or interview with a colleague or client, identify what types of reasoning and barriers to reasoning this person displays. Then think about why this person uses those types of reasoning, or is hindered by certain barriers to reasoning. You may need to consider corporate or team culture as a cause.
- Think of any "crutch" words or phrases that you overuse. By this, I mean the words and phrases that abound in corporate life and that we get into the habit of using unthinkingly. Stakeholder, utilize, leverage, lessons learned, impact…the list goes on. The reason I call them "crutch" words is that we lean on them when we feel tired or weak. But if we lean on them too much, the mental muscle of vocabulary – and choosing precise words to convey exact meanings – atrophies.
- Look at a recent document you drafted and use software to gauge readability. Where do you have the most trouble – overuse of passive voice, lengthy sentences, zombie nouns? Redraft two or three sentences to see how you could communicate your message more clearly by following the ABCs.
- Only once you have drafted a piece of writing, see what an LLM such as ChatGPT might produce on the same topic. Making sure that you don't include any sensitive or personal data, create a series of prompts in plain language on the topic you've written about. Compare the LLM output with what you have written. Is it better? Worse? Where are the gaps, assumptions, or hallucinations in the LLM-produced text? Where has it summarized more concisely?
- Find out if your organization has a corporate style guide. Sometimes this will simply cover aesthetics and branding: acceptable colors, fonts, and logos. Sometimes, though, it will be similar to an editorial style guide, setting out regional or national language preferences and giving advice on communicating clearly. Does it cover the ABCs? Does it give accurate and coherent advice about word choices? If not, or if you think the guide could be clearer, consider contacting the document owners and (diplomatically) offering to help when they next review the material.

- Look up the website/s relevant to your profession or activity. What advice and guidance do they provide about communications, reporting, and templates? Is it useful and relevant to your work? If so, can you share what you've found with colleagues, perhaps in a team meeting? This could be an excellent way to acquire continuous professional education or development (CPE/CPD) credits, as well as to improve professional practice in the team.

SUMMARY

- Before writing, we must think clearly. Understanding different types of reasoning – inductive, deductive, and abductive – can help us see where our critical thinking succeeds or fails.
- Barriers to reasoning are common to all of us. Bias and settled belief are often subconscious and therefore harder to detect. Confusing correlation with causation often leads to flawed conclusions.
- To write clearly, use the ABCs: active, brief, and concrete.
- Being active means favoring active over passive voice.
- Being brief means cutting unnecessary words and phrases.
- Being concrete means favoring verbs over zombie nouns. Too many zombie nouns in a sentence make it not only longer and more lumbering, but also more abstract and less engaging for readers.
- If 80% of your writing in English is active, it will be more readable. Sentences below 20 words will allow your readers to take a breath. Readability statistics will help you measure these.
- Large language models (LLMs) can help you draft material, with several caveats. Be wary of using any confidential company or personal data in prompts, and double-check any text an LLM produces for accuracy. As always, plain language will help you – in this case, in drafting prompts for the LLM.
- Using these principles and tools in everything you write, no matter how small, will quickly improve your confidence and competence. It will also make your other workplace tasks – including the review process – easier.
- Other hindrances to clear writing include personal hang-ups and corporatespeak, which you will have recognized from previous chapters.
- Mechanical mistakes can irritate or even mislead the reader. And it will give ammunition to the army of grammar snobs out there!
- Numerous resources can help you, including reference works (listed at the back of this book), professional bodies, trainers, and consultants. Choose what best suits your field, and make sure advisors have the required skills.

NOTES

1 The Institute of Internal Auditors Audit Executive Center, *2017 North American Pulse of Internal Audit. Courageous Leadership: Instilling Confidence from Within* (Lake Mary, Florida: The Institute of Internal Auditors, 2017), 26.

2 Joseph Holzman, "Current Regional Trends and Approaches to Critical Thinking," report from Goucher College micro-internship with Getting Words to Work®, January 2022.

3 *Journal of Accounting Education*, 21:3 (3rd Quarter 2003), 215–37. www.sciencedirect.com/science/article/abs/pii/S0748575103000277

4 Grant, 25.

5 I first explained the ABCs (and used the term "radical reporting") in a Chartered Institute of Internal Auditors event (June 2014). See also "Breaking Down the Audit Report," *Internal Auditor* (July 2021), https://iaonline.theiia.org/2021/Pages/Breaking-Down-the-Audit-Report.aspx

6 A note for the grammar-heads reading this. Yes, I could use terms like "subject" and "object," but that requires further clarification. After all, both sentences start with a subject. So I think it's easier to think in terms of doer/action/done-to

7 Orwell, "Politics," 119.

8 Sara I. James and Richard Chambers, "An Active Voice in Internal Audit Reports Inspires Action," AuditBoard blog, September 20, 2023. www.auditboard.com/blog/an-active-voice-in-internal-audit-inspires-action/

9 Thanks to Matthew Stibbe of Articulate Marketing for passing this tip on to me; he learned it from copywriter extraordinaire Clare Dodd. And copyeditor extraordinaire Benjamin Dreyer cites it in his excellent *Dreyer's English: An Utterly Correct Guide to Clarity and Style* (London: Arrow Books, 2020), 15. It also features in Yellowlees Douglas' *The Reader's Brain: How Neuroscience Can Make You a Better Writer* (Cambridge: Cambridge University Press, 2015), 37.

10 "Colson Whitehead's Rules for Writing" (2012) https://crhofmann.com/2012/08/27/colson-whiteheads-rules-for-writing/). You could say that he's taken more words than Jefferson to say the same thing, and you'd be right. But Whitehead is a master of prose in English, and he's also slipped in a reference to Arthur Quiller-Crouch's famous "murder your darlings" ("On the Art of Writing," *Lectures Delivered in the University of Cambridge*, 1913–1914, www.bartleby.com/190/), so he can do as he likes. (I think we can safely assume Whitehead's third sentence is tongue in cheek.)

11 Author of *The Oxford Guide to Plain English*, co-founder of the Plain English Campaign and founder of Plain Language Commission.

12 Oppenheimer, "Consequences."

13 Barbara Ehrenreich, *Bait and Switch: The Futile Pursuit of the Corporate Dream* (London: Granta Books, 2006), 28

14 "My English will never be 'perfect' – and that's what keeps a language alive," *The Guardian*, June 28, 2021. www.theguardian.com/commentisfree/2021/jun/28/english-perfect-language-arabic.

15 *Garner's Modern English Usage*. 4th ed. (Oxford: Oxford University Press, 2016), 983 (on which page I also learn that those of us taught to use the term "nominalization" are "jargonmongers"). Philip Collins discusses the pitfalls of nominalizations under "abstraction and abstract nouns" in *To Be Clear: A Style Guide for Business Writing* (London: Quercus, 2021), 53–54. See also Garner, *The Chicago Guide to Grammar, Usage, and Punctuation* (London:

University of Chicago Press, 2016), 451. I have used the term "verbal noun" interchangeably with "nominalization" for many years. According to Garner, though, "verbal noun" refers exclusively to gerunds. David Crystal agrees (*The Cambridge Encyclopedia of the English Language*. 2nd ed. [Cambridge: Cambridge University Press, 2003], 203), while dictionaries differ. Some define "verbal noun" as a gerund only, while others take a broader view: "**verbal noun** *n.* (formerly also †noun verbal) [after post-classical Latin *nomen verbale* (5th or 6th cent. in grammarians)] *Grammar* a noun formed from an inflection of a verb and partly sharing its constructions." (*Oxford English Dictionary*, www.oed.com/view/Entry/222360?redirectedFrom=verbal+noun#eid1222298520).

16 Cutts, *Oxford Guide*, 78.

17 It also features a grammatical mistake. "Needs" should be "need" as it refers to "retention and completion," two actions and therefore plural. But when sentences are this complex and confusing to read, it's not surprising errors creep in and no one then spots them. I'll discuss word-blindness in Chapter 9.

18 BP, Deepwater Horizon.

19 For a humorous example demonstrating the reverse, visit the US Government's Plain Language site's "Nine Easy Steps to Longer Sentences," www.plainlanguage.gov/resources/humor/nine-easy-steps-to-longer-sentences/

20 A. R. Moon, *Office English* (London: Ginn and Company Ltd., 1961), v.

21 Rogers and Lasky-Fink, 75.

22 Alee Marschke, "AI in IA: The use of artificial intelligence in Internal Auditing," report from Goucher College micro-internship with Getting Words to Work®, January 2024, 1.

23 Marschke, 4.

24 Lorelei Lingard, "Writing with ChatGPT: An Illustration of its Capacity, Limitations & Implications for Academic Writers," *Perspectives in Medical Education*, 12:1 (2023): 261–270. 269. www.researchgate.net/publication/372025489_Writing_with_ChatGPT_An_Illustration_of_its_Capacity_Limitations_Implications_for_Academic_Writers

25 IIA Global, Artificial Intelligence 101 Series: ChatGPT for Internal Auditors (2023), 14. www.theiia.org/en/content/tools/professional/2023/artificial-intelligence-101-series-chatgpt-for-internal-auditors/. See also Irena Ostojic, "On the Frontlines: The Potential, and Perils, of ChatGPT," IIA Global blog, October 23, 2023 https://internalauditor.theiia.org/en/voices/2023/on-the-frontlines-the-potential-and-perils-of-chatgpt/

26 www.theguardian.com/technology/2024/apr/30/us-newspaper-openai-lawsuit

27 Lingard, 263. See also Mignon Fogarty's AI Sidequests, including mention of "hallucination": https://ai-sidequest.beehiiv.com/p/ai-trustworthy

28 Young, "Garbage language." Philip Collins, however, would disagree with Young's statement that jargon is "actually purposeful." In *To Be Clear*, he says, "Genuine fields of scholarship, such as epidemiology or inorganic chemistry, have technical vocabularies for which there are no colloquial alternatives. Jargon, by contrast, is the deliberate translation of ordinary ideas into complicated word patterns in order to exclude the uninitiated. Business associates mimic their seniors to show that they belong. It makes a low art form sound like the highest science. But business is not a science. A business involves a complex array of human interactions. It is part history, part sociology and part anthropology. It is a realm not of scientific proof but of rhetorical proof." (31)

29 "U and Non-U: An Essay in Sociological Linguistics," *Noblesse Oblige*, ed. Nancy Mitford (Oxford: Oxford UP, 1956), 1–28. The other essays in the collection are funny as well as insightful – unless you are the kind of person who gets nervous about using the "wrong" term for items of furniture or meals. Trust me, life is too short.

30 "Being able to fluently use acronyms gives us a sense that we have mastered the strange linguistic games that fuel organisations. With proficient display of the arts of the acronym, people can feel that they are more skilled, knowledgeable and worthy of their position." (Spicer, *Business Bullshit*, 37). See also Turpin, Martin Harry, Mane Kara-Yakoubiam, Alexander C. Walker, Heather E. K. Walker, Jonathan A. Fugelsang, and Jennifer A. Stolz, "Bullshit Ability as an Honest Signal of Intelligence," *Evolutionary Psychology* (April–June, 2021): 1–10, https://journals.sagepub.com/doi/pdf/10.1177/14747049211000317

31 So I'll let Benjamin Dreyer do it for me: "The use of 'onboard' as a verb in place of 'familiarise' or 'integrate' is grotesque. It's bad enough when applied to policies; applied to new employees in place of the perfectly lovely word 'orient', it's worse. And it feels like a terribly short walk from onboarding a new employee to waterboarding one." (*Dreyer's English*, 157).

32 Oxford: Oxford University Press, 2020, 48–56.

33 Sawyer, Lawrence B. and Mortimer A. Dittenhofer, assisted by James H. Scheiner, *Sawyer's Internal Auditing: The Practice of Modern Internal Auditing*. 4th ed. (Altamonte Springs, Florida: The Institute of Internal Auditors, 1996), 781.

34 Garner, *Quack This Way*, 113.

35 Rob Reinalda, *Why Editors Drink* (Word Czar Media, 2020), 13.

36 From the peerless Capital Community College Guide to Grammar and Writing, https://guidetogrammar.org/grammar/modifiers.htm

37 Yellowlees Douglas discusses visual, speech, and auditory systems and their overlapping roles in helping us interpret both written and spoken communication (148–54).

38 Cecelia Watson, *Semicolon: How A Misunderstood Punctuation Mark Can Improve Your Writing, Enrich Your Reading and Even Change Your Life* (London: 4th Estate, 2019), 57–69, 71–89.

39 Elena Cresci, "Oxford comma helps drivers win dispute about overtime pay," *The Guardian*, March 16, 2017, www.theguardian.com/books/2017/mar/16/oxford-comma-helps-drivers-win-dispute-about-overtime-pay; Jon Sharman, "Oxford comma proves pivotal in delivery drivers' claim for overtime pay," *The Independent*, March 17, 2017, www.independent.co.uk/news/world/americas/oxford-comma-drivers-claim-overtime-pay-dispute-a7635991.html

40 Chris Stokel-Walker, "The commas that cost companies millions," *BBC Worklife*, July 23, 2018, www.bbc.com/worklife/article/20180723-the-commas-that-cost-companies-millions

41 David Shariatmadari, *Don't Believe a Word: From Myths to Misunderstandings – How Language Really Works* (London: Weidenfeld & Nicolson, 2019), 40. All of Chapter 1, "Language is Going to the Dogs," 21–40, is excellent on the persistence and ignorance of grammar snobs. See also Dreyer, "Rules and Non-Rules," *Dreyer's English*, 7–19; Lane Greene, "Is Language Logic?," *Talk on the Wild Side: The Untameable Nature of Language* (London: The Economist/Profile Books Ltd, 2018), 32–64; and John McWhorter, *Our Magnificent Bastard Tongue: The Untold History of English* (New York: Gotham Books, 2008), 63–64.

42 Guide to Grammar and Style, http://jacklynch.net/Writing/u.html
43 See Dreyer, 70–81. "A Note on Indian English" (73–75) will help anyone better appreciate the origins and richness of the English used by the second-largest English-speaking population in the world.
44 Cutts has an excellent chapter on inclusive language (*Oxford Guide*, 194–204), and Dreyer discusses it wittily and sensitively (*Dreyer's English*, 88–92). Readers with a more literary bent may enjoy Anne Fadiman's "The His'er Problem," *Ex Libris: Confessions of a Common Reader* (London: Penguin, 1998), 58–64. See also "What's in a word? How less-gendered language is faring across Europe," *The Guardian*, November 4, 2023. www.theguardian.com/world/2023/nov/04/whats-in-a-word-how-less-gendered-language-is-faring-across-europe
45 *Write It Right: A Little Blacklist of Literary Faults* (New York: Neale Publishing Company, 1909; repub. with a new introduction by Paul Dickson, Mineola, New York: Dover Publications, Inc., 2010), 49. And yet, until recently, it was completely acceptable to say "manageress" in the UK (and likely still in some regions)!
46 Chartered Accountants Ireland, *Technical Release 01 21: Reporting to Third Parties – Guidance for Accountants*, 2021, 12–13 www.charteredaccountants.ie/docs/default-source/technical-documents/technical-releases-alerts/tr-01_2021-reporting-to-third-parties.pdf?sfvrsn=df13aa7c_2
47 The Institute of Internal Auditors, The Global Internal Audit Standards™. Lake Mary, Florida: The Institute of Internal Auditors, 2024, 79, https://globalinternalauditstandards_2024january9_editable.pdf

Chapter 5

Planning

This chapter is as important as the previous ones, and complements them. If you are willing to analyze your thoughts rigorously and seek the best words to convey them, then it's only right to organize them logically.

You may think that your organization's suite of standard templates will help you, and they may. That is, they may guide you as to which information goes in which parts of the template. What they won't do is your thinking for you. That you must do yourself. Not only will it produce better reports, but it will make you feel more organized, confident, and in control of your subject matter. You'll be surprised by how many benefits there are to organizing your thoughts before writing – benefits you probably never suspected.

Planning what you will say and in what order is essential to communicating clearly. If you don't take time to plan, whatever emerges from your writing frenzy is likely to lack coherence and flow. Abraham Lincoln apparently said, "If I had five minutes to chop down a tree, I'd spend three minutes sharpening my axe." Whether he said it or not, the principle is one that will guide us throughout this chapter and beyond. The most important thing to remember is that the bigger the task and the tighter the deadline, the greater your need to plan.

No one ever congratulated a writer on hitting an (often arbitrary) deadline by producing something incoherent. But the inherent logic of this principle – if something's important enough to say soon, it's important to say it well – seems lost on most people. If you ask them what they would do if their house was on fire, they'd answer immediately: they'd call emergency services and shout "Fire!" Yet I've seen urgent incident reports, in the form of emails sent to board members in the middle of the night, that overlook this basic point. They should be the corporate equivalent of shouting "Fire!" – a few brief lines saying what and where the incident is, what is currently happening to manage it, and when readers can expect the next update.

One of these reports in email form ran to ten pages when printed out. Had there been a fire, the company's site would have burned to the ground by the time anyone finished reading the report.

This chapter will help you avoid this and other problems: losing your way in detail; not focusing on the most important points; and failing to see

DOI: 10.1201/9781003422365-7

important gaps and connections in your work. It will also suggest ways to share information in a timely and efficient way throughout investigations, reviews, and other engagements. This will, in turn, help you spend what time you have to write a final report more effectively.

We're returning to questions of time and discipline raised in earlier chapters. Whether reflecting on culture, choosing the right words, or forcing oneself to plan before writing, it can be stressful to hit pause when faced with a deadline. Taking a step back and using proven techniques will impose order on our thoughts and therefore our writing.

There are managers who say their teams lack discipline in writing. Often it's not exactly a lack of discipline at play. Most people don't realize they should plan their writing. After all, they've gathered all the information needed for their report and have an approved template to dump it in, don't they? So they charge in, filling the template with words and thoughts as they come, with predictable, usually unreadable results. An image that will remain with me forever comes from the American singer, actor, presenter, comedian, and activist Henry Rollins. At an Edinburgh Fringe event in 2012, he mentioned a US politician then in the headlines. "Whenever this man opens his mouth," Rollins said, "it's as though the words are trying to flee a fire inside his head."

I've read hundreds of reports exactly like that. Often they arise from simply not having the tools; sometimes, from fear or panic. We must always recognize our emotions when writing; if we are scribbling or typing hurriedly thinking only of a deadline, we must pause and reflect on a better approach.

Planning is that approach. Only through planning can you avoid presenting readers with reports that feature words, phrases, sentences, and paragraphs scrambling chaotically in every direction. Over decades of writing, academically and professionally, I have created a guideline for planning my work. Since I haven't seen it anywhere else, I'll claim it as mine and call it "The James Ratio." It's simple.

THE JAMES RATIO

Of the time you have set aside – or left – to write something, allocate 50% of that time to planning. This may seem excessive, but bear with me. Once we've studied the different tools and techniques you can experiment with and choose from, you'll see how usefully you can spend that 50% of your time.

You then need to spend only 20% of your time writing. This leaves 30% of your time for quality checks: spelling and grammar checks, using readability statistics, reading aloud, and peer or manager review. (If you have dyslexia, as mentioned in Chapter 4, you can use your 30% drawing on assistive technology and colleague support, rather than using spelling and grammar checks.[1])

The James Ratio – 50/20/30 – is something I've applied for several decades now, to school work, university essays, doctoral exams, and written assignments and portfolios for professional qualifications. In the workplace, I've found it valuable for everything: complex emails, letters to customers, business cases, and, yes, reports. It simply works. Emma Smith of Vodafone sees preparation as crucial to effective communication – bringing a presentation or paper to life for the reader. "Underpinning this is preparation. Rarely do well-prepared communications land badly."

Let me give you an example of one client who was extremely doubtful about my ratio. It was a small team of experienced and talented copywriters who had to produce website copy to tight deadlines daily. A typical production cycle would last as little as 30 minutes. Within less than half an hour of receiving the brief from the marketing team, the copywriting team of five would have to produce a clean piece of copy to appear immediately on the company's website.

The team worked swiftly and well, but always under pressure. Individual team members would get to work directly on copy for 15–20 minutes, then submit it to marketing for review. They would then have to make any required changes in as little as two or three minutes before submitting the final version for publication. Sometimes the marketing team would make changes themselves to save time, which demoralized the copywriting team. (It also led to the marketing team occasionally introducing typos, which then appeared on the company's website.) It was stressful.

When I introduced the copywriters to the concept of 50/20/30, they were stunned. One of them said, "OK, but if we receive a brief at 10:00am, we have to submit copy at 10:25 at the latest for publication at 10:30. Your ratio means we spend at ten to twelve minutes planning the content, five writing, maybe seven minutes reviewing. Can that work?"

I suggested they spend ten minutes planning, five writing, and five on peer review – a novelty for them. They could then aim to submit their copy to marketing at the 20-minute mark. They were still doubtful but promised to try it.

The following week, they started this process Monday morning. The benefits appeared within days: the team found the process smoother and less stressful. The marketing team requested fewer changes, if any. And a third benefit – discovered only weeks later – was that the process worked even when the team was understaffed. On one day, the manager told me, one member of staff was on holiday, another ill, and a third had to leave early to attend to his ill child. The remaining members of staff stuck to the new process – 50/20/30 – which helped them avoid panic, maintain focus, and deliver good copy.

I shared three techniques for organizing thoughts – planning – to this team, although there's one they preferred. The following sections explain all three and show how you can use them in different situations, no matter what kind of writing you do.

MIND-MAPPING

You may have learned how to mind map, and there are many useful resources. At its simplest, it's using shapes and possibly colors to represent thoughts, themes, and facts on a page. You can certainly use software to create mind maps, but there's nothing wrong with paper and pens – the old-fashioned way.

The point of a mind map is to spare you the confusion and pain associated with "brain dumping." I've met hundreds of people whose first step, when faced with a big report and looming deadline, is to "brain dump" – to spill onto the page every single fact and passing thought associated with the topic they can think of. This leads to one of two situations. The first harks back to the Henry Rollins quote: the report merely documents words fleeing the writer's head. The writer, losing focus, energy, and confidence, is then likely to give the draft report to a colleague to review. If this happens, the reviewer is likely to be as confused as the writer, and much more irritated. None of this produces good team morale or efficiency, let alone a readable report.

The second and less likely situation is that the writer, having committed the "brain dump," has the awareness to edit their own work. This is likely to involve removing 80–90% of what they've written and radically restructuring the rest. This is tedious, inefficient, and, again, demoralizing.

A mind map is the perfect tool for you to dump your brain. In a mind map – especially a hand-written one – you can be as confused, messy, and contradictory as you like. The point of the mind map is to thrash out all the elements of the topic you're working on. Every aspect, angle, and ramification can go in there. You can scrawl words, images, within shapes, with connecting lines, arrows, and question marks. It can be a work in progress, reflecting your own constantly evolving understanding of a topic.

In two separate cases, I worked with cybersecurity teams who had never mind-mapped before. This was part of training to help the teams – including, in one, the most senior IT leaders in the organization – communicate more clearly. The team members focused on different, highly specialized aspects of cybersecurity, and their resulting reports often duplicated information while overlooking common points. Mind-mapping, I thought, would help, and we had flipcharts and colored markers ready.

As usual in my training courses, I briefly showed attendees how to mind-map, divided them into two groups, and invited them to create two distinct mind maps, each on a different cybersecurity topic. When the time for the exercise was nearly up, I saw that each group had used at least two flipchart pages, and they were frenziedly filling more. In one case, the head of department, who was there and working with one group, said the activity was proving so useful that they should continue. I quickly revised my afternoon training plan and timings, and left them to it.

After 45 minutes, the groups brought their pages to the large board-style table we were all seated around. They placed their pages next to each other and taped them all together to create a monster mind map three pages across and two down. They then started drawing arrows among common themes, grouping points together within circles or rectangles, and adding more detail.

You'll appreciate they enjoyed the exercise. They proudly transported their monster mind map to their shared office and stuck it to the wall. Over the next months, it acquired further notes, shapes, and insights.

Both teams spent much longer on an exercise I'd originally planned 20 minutes for. What was the benefit? The immediate return on investment appeared during the session itself. In both cases, they realized that collectively they had been producing 10 to 12 reports a month, for different audiences. The mind map helped them see where these different audiences wanted broadly similar information – so the team reduced the number of reports immediately. This led to greater efficiency in producing the reports, and greater consistency across them. No longer would people receive three, four, or five reports on cybersecurity and have to ask what the slight differences among them meant. The data and messages were more coherent.

The less immediate return on investment was greater awareness of what different team members were doing. Although they all belonged to one main team, their specialized focus meant they often forgot or ignored relevant information outside their smaller teams. The mind-mapping exercise helped them realize how they could share, use, and expand on information across teams much more effectively. This in turn helped them produce fewer but more useful reports.

Let's return to the point about numbers and types of reports. If you are in a team or department that produces numerous reports to different audiences on the same topic (enterprise risk management or IT security, for example), this kind of exercise could be useful. If your team mind maps all the different reports and briefings, it's easier to see where you can streamline your work. If you can then reduce the number of reports you have to provide, but increase their quality, that alone would justify the time spent mind-mapping.

Even if you are producing only one report, consider the requirements of the different readers, both internal and external. A mind map can help with this, allowing you to sketch out readers as disparate as operational managers, board members, regulators, and even members of the public (for public-sector organizations who must publish reports or respond to freedom-of-information requests). Gathering all these groups, their expectations, and needs on one page can help you see where you can group together information relevant to all of them, which reduces repetition in your report.

It will also help reassure you that you are factoring in the relevant industry standards for reporting.

Emma Smith lists the various sources of information and guidance her teams use for reporting.

> For reporting we typically don't follow a framework. We use ISO and our own risk assessment behind our control framework. We have used NACD and other research into board reporting. Regular reporting follows a method I have used for some time: threat and risk (internal and external), where are we against where we want to be, how does that compare with other companies, and the plans to improve. We have key risk indicator metrics that measure the controls most crucial to mitigating risk. I collaborate with peers through industry forums (I4, ESAF, ACF, etc.) and we also commission our own benchmarking.

This list alone shows how much teams can benefit from a graphic representation of everything feeding into their reporting.

Another way in which mind maps can help plan is with complex projects requiring different communications. The example below comes from a coaching client, a senior financial crime manager, who led a team producing a new fraud leaflet. The mind map allowed her to see more clearly not only the readers but indeed the whole project.

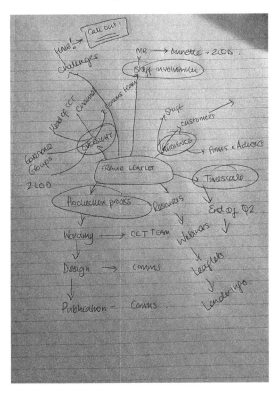

Whenever you face a complex project or report, mind-mapping can help spread out all the information at a glance. Having said that, many people don't take naturally to mind-mapping – I'm one of them. I instinctively prefer more ordered and linear ways of organizing my thoughts, and I'll share them with you shortly.

However, there is an advantage to mind-mapping if it's outside your comfort zone: you see things differently. Often when we do what is habitual and familiar, we risk skimming over information or making assumptions. By forcing ourselves to try a different approach, we force our brains to work differently. It can be uncomfortable, but extremely useful. As I said, I'm not a natural mind mapper, but when I've used it, I've benefited greatly.

Initially, I used mind maps for findings, and Chapter 7 will share that technique. But over the years, I've learned from students across the world – in France, Scotland, and Singapore, to name but three places – how they use mind maps to produce better-written communications. I've also worked closely with heads of internal audit and risk on mind-mapping the risks facing their organizations. They can then produce relevant, coherent annual or other regular plans for the board and senior committees to scrutinize.

Let's look at how this technique can help you organize your thinking at all levels, using a (fictitious) UK food manufacturer as our case study. In the first mind map, we see a few of the areas of internal and external risk facing this manufacturer. Whether you're in audit, risk, compliance, or IT security, you'll recognize these risks and be able to add many more, as well as see how this could help create a comprehensive yet focused operational plan.

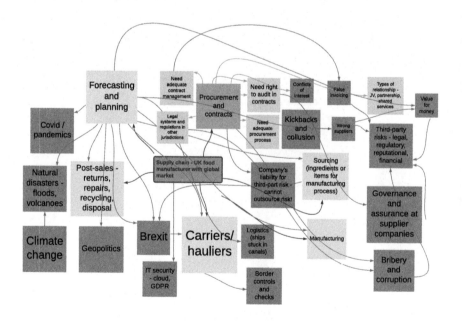

Now let's imagine you're an internal auditor reviewing contract management for that same UK food manufacturer. You'll have seen from the high-level mind map above that contracts feature, so it's important to assess the controls in place to mitigate contracts-related risks. You might come up with another high-level mind map, this time setting out the risks specific to contract management. From this mind map, you could create terms of reference, risk assessment documents, and other written communications essential to your engagement.

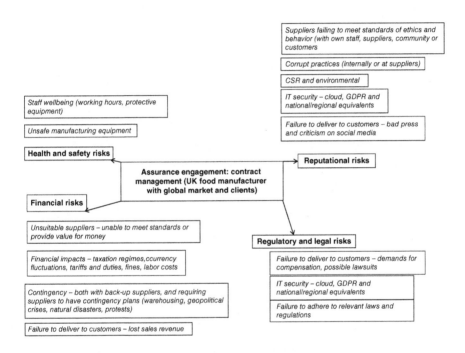

The third mind map is inspired by an internal auditor I met in Singapore, where high-school students learn mind-mapping as an essential study skill. This internal auditor explained that she uses mind maps to plan her testing for each engagement. This was new to me, so I asked her to tell me more. Apparently, she spends two to three hours before every audit mind-mapping numerous elements: risks, controls, expected statistical sampling methodology, existing data she can rely on rather than duplicate, and even external recipients of her test results.

What she meant by this was that, rather than planning to test only what was necessary for the immediate audit, she checked whether other parties required test data in the coming 6–12 weeks. Her reasoning was that if, for example, the risk function was requiring similar data in two months' time, why not find out their requirements now and share the results, rather than re-perform the testing? If the external auditors are planning to check the same controls but using slightly different criteria – again, find out their requirements and factor them into the same batch of testing. Internal and external assurance functions – internal audit, risk, compliance, external audit, and even regulators – often place reliance on each other's work (after assessing it's to the required standard). It therefore makes sense to find out where there are similar requirements so as to avoid duplicating someone else's work unnecessarily, or repeating work you've done recently. The mind map below shows how that fictitious internal auditor reviewing contract management can plan testing – what to test, what to rely on (and question), and for whom.

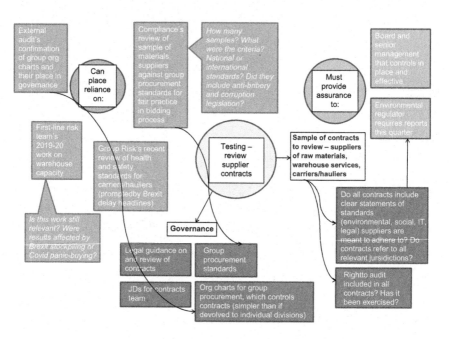

As with the cybersecurity teams with their monster mind maps, we have to ask about return on investment. Drawing shapes is all very amusing, but what is the tangible benefit? The Singaporean internal auditor quantified it starkly: "For every two to three hours I spend mind-mapping, I save myself two to three days of testing."

How does this aid reporting? First, anything that makes you work more effectively and efficiently frees up your time and energy, which you'll need to document your work. Second, reports will be more powerful if they communicate a coherent message. Rather than testing and reporting piecemeal, you can produce reports that reassure boards and regulators that you have considered "the big picture." Taking into account broader considerations such as organizational risk appetite, regulatory requirements, industry standards, and relevant legislation makes reports compelling calls to action, rather than a list of apparently low-level tasks.

After all, what we call internal and external assurance functions must perform many of the same tasks for the same reasons – but in slightly different ways. Mark Carawan of New York University has compelling insights into this dynamic. He's seen his fair share of reports, and knows better than most how many rivers of text must join together to produce a flow of useful information!

To take one example, Sarbanes–Oxley legislation affected how different assurance functions operated as well as what they had to report.

> For US SEC listed companies, the requirements for Sarbanes–Oxley (SOX) compliance and the formation of the Public Company Accounting Oversight Board had a profound impact on the external auditing profession and the roles performed by external auditors. Likewise, many SEC-listed companies drew upon the resources and skills of the internal auditors within their organizations to undertake work to assist in dealing with SOX, potentially diverting and diluting the risk-based internal audit roles within the organization. SOX likewise expanded the responsibilities of the compliance officers in such companies. So, in one piece of legislation, each profession was affected in different ways – impacting how each role interacts with the other two: from degrees of reliance (or potentially none at all), to the nature and character of co-ordination, collaboration, and communication.

Although Carawan describes a complex situation clearly, we can see how factoring these different elements into a report would be daunting. Mind maps can thus help clarify all the elements, sub-elements, and the numerous relationships among all of them. Think of it this way: if a board member asked you to produce a brief report about a topic as complex as this, you'd probably struggle to draft something coherent directly on the page. A mind map, though, would help you at least organize your thoughts before deciding what to write.

Louise McKay, quoted in previous chapters, is articulate on how moving from one specialist area to another requires a fresh perspective. As she says,

> internal audit reporting tends to focus on a much narrower area than risk reporting. Internal audit work usually focuses on something quite specific – a product, process or theme – whereas our risk reporting focuses on risk categories which apply across most or all areas of our organization. The challenge is to assimilate multiple data points to form an overall conclusion. That can be really tough, not least because there's often a degree of subjectivity and sometimes different data points tell you a different story! We also provide a view on the risk outlook for the organizations which is a key difference from audit reporting.

If you, too, need to have a broader perspective across your organization, a mind map will help you. Sketching out different areas, different risks, and different data points will almost always make it easier to analyze – and to produce a clear, coherent, meaningful report.

OUTLINES

You may heave a sigh of relief at this point. If you remember outlines from middle or high school, you'll recall how structured and linear they are – the opposite of mind maps. For those of us with a more analytical, less creative bent, outlines come naturally. Many people who've never encountered them before embrace the logical progression of a classic linear outline.

An outline is simply a way of organizing your thoughts and data hierarchically. In Chapter 1, I quoted a German banker who said that his UK colleagues seemed to write "thought and word salads," whether in emails, briefings, or reports. This was almost entirely due to differences in education: whereas teaching children to use outlines is standard in most western schools, this is generally not the case in the UK and the Republic of Ireland.[2] The result is people learning far too late – if at all – one of the best techniques for organizing and ordering thoughts and facts before writing in earnest.

If you've never produced an outline before, there are plenty of resources available, especially online. The excellent Capital Community College Guide to Grammar and Writing has a page on outlining, as well as one on mind-mapping, which it refers to as "clustering".[3] There's even an outline function in Microsoft Word, although it's almost always better to write your own.

In a classic linear outline, for a school essay, for instance, you'd organize your research and insights hierarchically. You would represent the hierarchy using a combination of Roman numerals, capital letters, Arabic numerals, and lower-case letters, depending on how many levels or layers of information you go into within each section.

At the most basic level, an outline looks like this:

  I. Introduction
    A.
    B.
    C.
 II. First point
    A.
    B.
    C.
III. Second point
    A.
    B.
    C.

IV. Third point
   A.
   B.
   C.
V. Conclusion
   A.
   B.
   C.

One important principle when developing an outline is that if you have only one sub-section under each main heading, it may not be able to stand on its own. So, taking the example above, if we have:

  I. Introduction
    A.
    B.
    C.
 II. First point
    A.
    B.
    C.
III. Second point
    A.
IV. Third point
    A.
 V. Conclusion
    A.
    B.
    C.

This would suggest that the second and third points are weaker than the first. Maybe there is less evidence for them; maybe they fall into different categories. If this is the case, analyze why your outline appears unbalanced. Chances are you will realize how you can strengthen the weaker points, or discard them altogether.

I've used introduction and conclusion to reflect typical essay content. In a report, however, you would probably start with the executive summary, and the final element would depend on your report. In some teams, the recommendations or next steps might appear here. In others, appendices will close the report outline.

Let's look at turning a mind map above into an outline. If we look at the first mind map – the one about the fraud leaflet project – we can immediately see distinct main points. So a first outline may look like this:

I. Fraud leaflet project
  A. Audience includes internal and external readers
  B. Requires oversight from various internal teams
  C. Production (including timescales and resources)
II. Audience
  A. Customers
  B. Firms
  C. Advisors
  D. Staff
  E. Oversight
  F. Second line of defense (risk)
  G. Governance groups
  H. Head of CCT
  I. Channel
  J. Comms team
III. Production
  A. Timescales – due by end of Q2 (time challenges?)
  1. Production stages and assignments
    a. Wording – CCT Team
    b. Design – Comms
    c. Publication – Comms
    d. Staff involvement – "me" + another team member
  2. Resources
    a. Webinars
    b. Leaflets
    c. Lender info

You could then develop such an outline further, writing a sentence or two for each element. This is particularly helpful when you dread writing and will do anything to avoid beginning. If you say to yourself, "I'm just going to start scribbling a mind map, then an outline," you'll have started the most important part. Once you've done that, writing is much easier – you simply start at the beginning, go stage by stage through every element until the end, with greater focus and confidence. Because you have already thought through what belongs in your report and in which order, you will write less but say more. Your report will include what is not only true but also relevant, as the outline reduces the temptation of including random facts and observations. This is why if you spend half your time planning, you drastically reduce the time spent writing.

In Chapter 8, you'll see a more developed outline, reflecting the content readers will find most useful in a typical report. Don't skip to it now, though – there's more work to do between now and then, so that you can better appreciate and use the "full" outline.

I first learned to use an outline when I was 12 or 13. At that time, and for several years afterward, it was common practice for teachers to assign an essay due several days later, and to require an outline partway through. For example, if the teacher told the class Monday their essays would be due the following Monday, he or she would ask to see a detailed outline by Wednesday. This then allowed the teacher to check that students had understood the topic and were making progress in their research, analysis, and thinking. More importantly, if a student hadn't understood the assignment, or had veered off on a tangent, it was easier for the teacher to identify the error in a one-page outline, thereby allowing the student time to correct course. This reflects the James Ratio – half the time allotted to the assignment was spent on the outline, with the remainder on writing, reviewing, and editing.

Decades later, I discovered how useful outlines could be throughout research, an investigation, or audit fieldwork. They provide a handy summary of key points, ordered hierarchically, to discuss with colleagues or managers. Again, if there are any misunderstandings or disagreements, it's far easier to correct an outline than to redraft a 10-page or longer report. It's also less stressful, as changes take place well ahead of reporting deadlines. A crucial point to remember is one made by David Foster Wallace, speaking about argumentation. Since this term refers to making a case using logical reasoning and supporting evidence, it includes reporting. "If you couldn't, if forced, put it into an outline form, you're in trouble."[4]

Another advantage I discovered accidentally. Once, during a high-profile internal audit, I was preparing for the first regular "checkpoint" meeting with the head of the area being audited. (Other teams may call these meetings "tollgates," or yet different terms.) This person had not experienced risk-based internal auditing and associated an audit with the purely accountancy-based work he'd observed many years before. Since this area hadn't been audited in over ten years, there was apprehension about what we might find, and this senior manager was both nervous and defensive. This in turn made me slightly nervous about how the first meeting would go, so I organized my thoughts in a simple, handwritten outline. This was more of an aide-mémoire for my use, so I didn't bother to type it up or make it look formal.

When the meeting began, the head of the area was sitting stiffly across from me, behind his imposing desk. I took out my handwritten outline and mentioned that this was simply some notes I'd made about what the testing showed to date. To my amazement, he swung his chair around the table to sit next to me and started poring over the outline. A former academic, he probably immediately recognized the structure. The fact it was handwritten made him relax – after all, it wasn't a formal document to which he'd have to respond.

We then proceeded to have one of the most open and fruitful conversations I've ever had with an audit client. Studying the outline, he suggested grouping two distinct points together, as he felt they shared a root cause. This then prompted a discussion about underlying problems with IT systems, and how it was affecting operations across the organization – useful information for anyone to have. Most startlingly, he suggested that my third point was the most important one. He explained his thinking, sharing frankly with me why he felt it posed the greatest risk to his area.

You can imagine how surprised I was that a simple (handwritten) outline had opened the floodgates to so much useful information. What's more, this person, who had previously been defensive and closed, showed himself willing to be frank and transparent, including about topics I hadn't even suspected.

Over the following weeks, I used the same technique on other occasions. Each time, the outline prompted greater openness, more information-sharing, and useful insights. It also helped me gradually develop, outline by outline, what would eventually become the draft report. Because I had been regularly discussing the outlines in their various versions with the head of the area, there were no surprises by the time the draft report came out. I had been able to raise sensitive topics, seek his input, and craft clear, accurate statements of the problems the area faced.

It was a far cry from the typical experience of having a final meeting (whether you call it "close-out," "end-of-fieldwork," "concluding," or something else). In these meetings, it's common for the auditors or investigators to discuss what they have found, even to present a few bullet points on slides. Typically, everyone agrees enthusiastically to what they've said – and then expresses shock and dismay when they see the report, putting into writing what everyone agreed verbally. You can feel impatience with this response – after all, most teams issue written reports. How on earth did the clients, team being audited, or area under review *think* we would convey our findings – through cave painting, or interpretive dance?

But this overlooks the human element. Seeing something written down, whether a report suggesting improvements, a disappointing exam grade, or a medical diagnosis, often comes as a shock. As long as things remain unwritten, we can kid ourselves that they may change, or the severity of the message may lessen. Once we see the evidence on paper (or screen), however, denial is difficult, and defensiveness common.

So outlines can improve not only how you organize your thoughts, but also how you communicate with colleagues and clients before drafting a report. It can make much easier the process of gathering and analyzing information, and gaining consensus.

This is why I've used handwritten outlines ever since. The astonishing thing is that not once has an audit client reacted badly to it. Now, certain cultures – organizational and other – may balk at seeing anything handwritten. But my experience is that this has simply communicated to the client, "Look, this is just you and me. I'm sharing my thoughts with you, and they reflect what the current fieldwork results tell me. What do you think? What's missing? What doesn't look right? What seems like an anomaly?" Every time, the audit client has entered into the discussion in a spirit of openness and cooperation. I hope you experience this, too.

The last point I want to make about outlines is my emphasis on writing them by hand. There may be many excellent reasons why you won't want to do this, or share the result with clients. Perhaps you have dyslexia, and writing is difficult enough, or difficulties physically using pen and paper. Maybe, as mentioned earlier, clients will respond badly to the sight of a handwritten document.

However, if possible, try writing by hand at least occasionally, even if only for your own benefit. The process of writing by hand is physically and mentally different from typing, as the novelist Fay Weldon advises. "Develop the manual techniques of writing," she writes, "so that as the mind works the hand moves."[5] This may seem unnecessarily old-fashioned to you, even alien, but I strongly recommend it. As well as learning how your brain works differently when writing and when typing – much as mind-mapping may shift your perspective – you'll also hone the skill.

Nicholas Carr's book *The Shallows: What the Internet Is Doing to Our Brains* is excellent on the different ways in which we read and process information.[6] You don't have to be a technophobe to be aware that, as life becomes ever more automated, our conscious and unconscious habits change. As a member of the pre-computing generation, I reached voting age before ever using a computer to write an essay or article. The habit of using keyboards, tablets, and smart devices has changed the way I think and certainly write. This is why I try to write mind maps, outlines, and notes by hand regularly. If we keep practicing older tools and techniques, we will be able to retain them, to have a choice, and to not be dependent on only one mode of thinking or writing.

Mini-outlines

These are exactly what they sound like – a rapid-fire way to seize the key elements of what you need to communicate. Some might say they're not even outlines, as there's little hierarchy other than listing items in order of importance. It's simply a matter of challenging yourself to summarize your investigation, proposal, or update in no more than six bulleted items.

These are more useful for shorter documents: mini-reports, interim memos or briefings, even emails. I've used them for ad hoc investigations when a site incident required a speedy report within days. After gathering information from sources of varying insight and reliability, I'd find it best to sit down and write (again, by hand), what I knew in six points. For example, if the board asked you to report on a major IT failure, the high-level mini-outline might look like this:

1. The X server failed last night.
2. We've lost £15 million in transactions as a result.
3. If we don't repair or replace within X days, we'll lose another £50 million.
4. This is the quickest and best way to replace it.
5. This is the main objection to replacing it. [Note – this will usually come from the senior manager who hasn't repaired or replaced the server in years and is trying to block anything that makes his or her negligence clear.]
6. This is why we should ignore this objection.

Would your final report to the board look like this? Of course not – but sketching out in a few lines the main, unavoidable points to include will give you the spine of your report. You can then flesh out where necessary, without digressing or going into unnecessary detail.

Mini-outlines are also excellent for important emails. If you find that you are receiving late, inaccurate, or no responses to emails you send, consider using mini-outlines. I used to work with a client's actuarial and IT teams to produce a monthly newsletter about the EU financial directive Solvency II. Thrilling stuff! This meant extracting regular updates from people not known for their keenness to interact with others. The internal communications team had previously written the newsletters and had had difficulty getting the right information in good time from relevant experts. What they received was often too detailed, too technical, or not immediately relevant.

Using mini-outlines to hone my call-for-information emails helped reduce the time and effort everyone spent. First I created my mini-outline of information that *had* to appear in the email.

1. Solvency II Newsletter Issue X [insert month and year]
2. Updates needed by [deadline date]
3. Must include information about new Solvency II-related projects or systems launching this month
4. Must include information about major changes or delays to existing Solvency II-related projects or systems
5. If no change since previous issue, please say so
6. Word limit for updates

Looking at the six items, 1 and 2 could appear in the subject heading of the email. This automatically reduces the amount of text in the body of the email. The resulting email ran like this:

Subject: Newsletter Issue 13 – call for updates (deadline 5 August)

*Hello – yes, it's that time again. Can you please send me **250 words maximum by 5pm this Friday, 5th August**, about progress in your area since mid-June? The focus should be on what changes people in the Group have seen, or will see between now and mid-October. Issue 12 attached for reference.*

Thanks in advance. As always, please get in touch if you have any questions.

Sara

[contact details]

The result was that people responded promptly with concise, relevant updates. When I share this approach with training and consultancy clients, they are astounded by the benefits. One cybersecurity expert said she'd received relevant responses by her deadline for the first time ever using the mini-outline technique.

There are other ways of refining your approach to improve communications. You may need to occasionally call people to check their workload and let them know the email is arriving. Creating this rapport of understanding makes it likelier they will try to deliver what you need. If it's impossible, then at least you'll have been warned and can plan around it.

Giving people a get-out clause for 5 can also promote efficiency: "If I don't hear from you by [deadline date], I'll assume there's no change since your last update." This tells them that it's fine to have no news, and also that they don't need to go to any effort to convey this to you.

You can also invite people to tell you if they are no longer responsible for this information, but that's likelier to happen when reporting takes place at greater intervals. Weekly or monthly updates mean your correspondents are likelier to inform you of any staff or remit changes.

Mini-outlines also work well for mini-reports. Many teams are moving – whether under the guise of "agile" methodology or prompted by changes to working from the Covid-19 pandemic – to shorter, sometimes one-page reports. And a mini-outline is perfect for working out exactly what you must include in that one page. If you have so little time to make an impression, you'll want to use every word and line wisely.

Paul Breach, Head of Audit Practices and Operations, Internal Audit at LSEG (London Stock Exchange Group), describes how reporting practices have changed recently, and the results for clients.

The incremental "reporting" aspect provided auditors with the templates and expectations to discuss and report conclusions throughout the duration of the audit, before wrapping it all together at the end. This approach to communicating "as we go," rather than waiting to the end, was particularly well received by stakeholders.

If this is an approach your team would benefit from, think of how mind-mapping, outlines, and mini-mind maps can help you refine your thinking so as to present readers with something as clear, focused, and useful as it is concise.

Try all three techniques and see which works best for you and in which circumstances. You may find that mind-mapping is a boon during cross-divisional reviews, or benchmarking exercises. Outlines may give you comfort – and your manager's approval – before writing a report. And a mini-outline may help you craft emails and updates that hit the target every time.

A final set of planning images shows the return on investment for planning. These come from another coaching client, an IT auditor whose work was of the highest standard but who lacked confidence in organizing her thoughts. I shared these three techniques with her and asked her to try each for her next report. She came to the next coaching session with the news that she hadn't had to write a report in the previous week. However, she was starting an IT integration review and had to produce terms of reference for it. She tried all three techniques – by hand – with the following results:

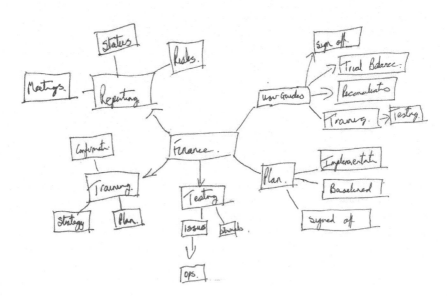

Mind map.

Outline (in spirit if not in exact form).

Mini-outline and...???

Yes – what *is* that final image, below the mini-outline? It looks like columns of information, linked across the top by the big themes (governance, methodology, RAID, resources and training, procedures). When asked, the client explained that once she'd started experimenting with the different techniques, her brain started going into overdrive, thinking of all the different aspects of this work and how she could group them. So she came up with these four rather than three representations of what she would have to consider in her terms of reference.

Then came the "So what?" moment: what was the return on investment? Well, she said, she spent 45 minutes at home one evening scribbling away happily. The next morning, she went into the office and by 10:00 a.m. had produced not only the review terms of reference, but also her risk and control assessment document. By 11:00 a.m. her line manager had signed off both, perfectly happy.

Using these techniques not only saved her time, but also pain. For the first time, she went into the office without dreading the moment she would face a blank page or screen and start writing. She had done the hard part – the rigorous analysis and organization of her information – and the next morning was eager to document her plans and insights.

She was ready to write.

I hope this chapter has given you plenty of examples to start experimenting. Each technique can be used for many different purposes. Mind-mapping can produce annual plans; investigation, review, or testing plans; terms of reference; risk and control documents; and, as you'll see in Chapter 7, insights into root cause.

Outlines can be aides-mémoires for regular meetings, as well as a useful tool to agree report content with managers and other reviewers. Once you've written an outline, the writing "proper" comes much more easily. Finally, the mini-outline of six bullet points will help you with emails, interim reports, briefings, and memos.

You may find that you feel much happier and more confident when writing once you have your preferred planning tool or technique. It's not surprising – after all, you wouldn't build a house without having architectural blueprints first, would you? Moreover, any task is easier when you know exactly what your next step is – writing is no different, as even Ernest Hemingway said. "I always worked until I had something done and I always stopped when I knew what was going to happen next. That way I could be sure of going on the next day."[7]

At this point, some of you may be thinking of Chapter 4, which discussed the role of LLMs in drafting text. Why not use software to brainstorm, to organize thoughts before writing? There is of course no reason not to – provided you keep in mind the security and copyright elements mentioned earlier, and maintain control of the inputs and outputs.[8]

For Alex Psarras, Associate Director, Internal Audit and Financial Advisory at Protiviti UK, LLMs offer tremendous benefits to those working in audit, risk, compliance and IT security – who must also be cognizant of the risks.

> For me, the obvious opportunity is to allow assurance professionals to ingest large volumes of information, to get quicker insights or generate reporting. This could be automating manual activities like summarising multiple documents, emails or chat threads to get to the key points, translating technical documentation, diagrams or text in foreign languages into plain English, providing real-time meeting feedback, and creating the first draft of a report based on all information known so far.

However, he also points to downsides – the greatest risks being "biases in the training data, the overreliance on LLMs without human oversight; and the potential for misinformation and hallucinations," as mentioned previously.[9]

Alexander Rühle agrees with Psarras on the efficiency LLMs can bring to an organization using them thoughtfully. Not only can they help us "get rid of low-level work, unloved and of no value in times of ever-increasing documentation requirements," they can change functions fundamentally, and for the better. From an internal audit perspective, he writes,

> I claim that AI is the greatest opportunity for us to move towards transformative auditing. What is necessary? The new standards require internal audit governance changes. That is an opportunity to rethink the company-specific purpose based on internal audit's purpose, to define the how and what – the strategy.

Returning to the James Ratio, we have covered the 50% of the time spent on planning, reducing writing to 20%. Chapter 4 covered the ABCs and other ways to make what you write clear, concise, and meaningful. The remaining 30% is for review – reviewing your own work, and having others review it. We've already mentioned spelling and grammar checks, readability statistics, and reading out loud. Chapter 9 will focus on the actual review process in teams, and how to make both peer review and manager review more positive and productive.

In the next chapter, we will move from planning to structure. Once you have gathered and ordered your information and insights, it's much easier to structure reports. However, Chapter 6 will help you understand traditional report structure, the pros and cons of standard templates, and the usefulness (or otherwise) of automated reporting.

ACTIVITIES

- Start with a mini-outline. Think of an important email you sent recently, but don't look at it. From memory alone, write a mini-outline – no more than six bullet points, each one no longer than a sentence. Now compare it to the email you actually sent. What are the differences? Was your email longer or shorter? Did it put the most important information first, or bury it in detail? Most importantly, did you get the answers you needed to your email on time? If not, maybe your readers got lost.
- Once you are confident in your own ability to create outline, ask an LLM to create one for a particular topic. Make sure you don't use any organization- or person-specific data in the prompt, though. For example, you could ask "Please create an outline for testing outward payments in a manufacturing company," or "Please write an outline of how I can check basic cybersecurity controls in an insurance company."
- Mind map a current project – this can be an upcoming investigation or review, or a piece of research for work or study. What connections have appeared using the mind map? Do some aspects of the project or topic have more links than others? Are there any gaps you hadn't realized?
- When it's time for you to prepare writing up your project, select the most relevant parts of the mind map and create an outline from them. Share it with your manager, reviewer, or supervisor. Are you on the right track and ready to start writing? Or do you need to consider some elements further?

SUMMARY

- "If I had five minutes to chop down a tree, I'd spend three minutes sharpening my axe." (*attrib.* Abraham Lincoln)
- The more you plan, the less – but better – you will write. Try the James Ratio: 50% of your report-writing time on planning, 20% on writing, and 30% on quality checks.
- Rigorously thinking through all the elements of your topic will enable you to see the complexities, connections, and gaps more clearly. You can then develop annual plans, terms of reference, research papers, reports, and more with confidence.

- Outlines – such as those learned in school – have much to offer in the workplace. Whether you are gathering your thoughts for a client meeting, sketching out the order of a draft report, or even planning an important email, outlines help you order the main points and supporting evidence logically. As with drafting, judicious use of LLMs can help create an initial outline for you to review and develop further.
- The more you experiment with different planning tools and techniques, the greater your flexibility. They will help you think differently about the information you want to convey, and consider how to appeal to a wide range of readers, whether in one report or several.

NOTES

1 There are websites in both North America and the UK, as well as other countries, offering support and advice for people with dyslexia. These include The International Dyslexia Foundation (https://dyslexiaida.org/), The American Dyslexia Association www.dyslexia.me/), Dyslexia Canada (www.dyslexiacanada.org/), The British Dyslexia Association (www.bdadyslexia.org.uk/), Dyslexia UK (www.dyslexia.uk.com/), and The Dyslexia Association (www.dyslexia.uk.net/).

2 George Orwell would likely have attributed this to what he called "the lack of philosophical faculty, the absence in nearly all Englishmen of any need for an ordered system of thought or even for the use of logic." "The Lion and The Unicorn: Socialism and the English Genius," *Why I Write* (London: Penguin Books, 2004), 11–94.

3 Capital Community College Guide to Grammar and Writing, www.guidetogrammar.org/grammar/composition/brainstorm_outline.htm and www.guidetogrammar.org/grammar/composition/brainstorm_clustering.htm.

4 Garner, *Quack This Way*, 78.

5 *Letters to Alice on First Reading Jane Austen* (New York: Carroll & Graf Publishers, Inc., 1990), 30.

6 London: W. W. Norton & Company, 2011.

7 *A Moveable Feast*. The Restored Edition (London: Arrow Books, 2010), 22.

8 For a practical demonstration, see Jon Taber, Abbas Al Lawati and Alex Rusate, "The Softball, The Hardball, and the Curveball," Audit 15 FUN podcast, January 24, 2024. https://jontaber.substack.com/p/the-softball-the-hardball-and-the

9 See also www.theguardian.com/technology/2018/oct/10/amazon-hiring-ai-gender-bias-recruiting-engine

Chapter 6

Structure and layout

In the previous chapter, we laid some of the groundwork for this one. Once you have carefully thought through your topic, research, and insights, and organized them in a planning document, it's much easier to structure your report logically.

However, do you even need to produce a report? Perhaps this is the most radical idea of all – a book about reporting that questions reports. It's a question worth asking, though – think of how many different techniques and media are available to you, and how your readers access information. They probably use screens, but this doesn't mean continuing to produce Word documents for them to scroll though. They may prefer a presentation, or video, or a simple email – we'll cover all these possibilities in this chapter. Whatever medium (or media) you use, though, you must have a logical, reader-friendly structure and layout.

At this point, you may be thinking of standard templates you must use in the workplace, and we'll cover their pros and cons later in this chapter. For now, though, this chapter may help you better use your templates, or even suggest revising them – so it's worth reading through.

We will start with report structure, then cover layout (the "look and feel" of the document on page or screen). The layout discussion will also address making documents accessible. By improving how they appear, you can help all your readers – not only those you assume will benefit. Finally, we will talk about templates, including automated reports. Sample templates throughout the chapter should prompt you to think creatively about how best to present your report content.

STRUCTURE

The structure of any report should be as simple as possible. However much information you feel you need to include, a simple structure will increase the chances your readers will keep reading and benefit from the detail. All too often, though, reports meander through not only dozens or hundreds

 DOI: 10.1201/9781003422365-8

of pages, but numerous complex headings, sub-headings, sub-sub-headings, sub-sub-sub-headings...you get the idea.

The result, far from "signposting" the content clearly, has the opposite effect, as Steve Pressfield describes.

> There's a story about an embassy that was sent once to the ancient Spartans. The foreign envoys spoke for hours before the assembled citizens, seeking their aid. When they had finished, the Spartans declared, "We can't remember what you said at the beginning, we were confused by what you said in the middle, and by the end we were all sound asleep."[1]

Many people think they can avoid this by just the sort of "signposting" mentioned above. By creating lots of headings, sub-headings, etc., throughout the report, they hope to remind the reader of where he or she is, whether on page 4, 19, or 72. However, it usually doesn't work because it doesn't address the problem. Whenever I've asked someone why there are multiple levels of headings in the report, the response is always the same: "To make it clear to the reader what I'm saying."

This should raise an obvious question. If you know that what you're saying isn't clear, how will adding more text (italicized, underlined, in bold, at various intervals) help? The solution is to *fix the writing that's already there.* If you find yourself in this situation, go back to the first five chapters – you have to start with critical thinking and plain language before thinking of how to present it on the page. Otherwise, you will be like someone fussing over the wallpaper in an unbuilt house.

There is another analogy for this tendency to use complex report structures to impose order on chaotic writing. Imagine a badly paved road, filled with potholes and fissures. The obvious solution is to re-pave the road, making it fit for people to travel on safely and smoothly. However, if you believe the solution is to leave the road as it is, but add lots of signposts along the side of the road, don't be surprised if people don't want to travel along it! Similarly, a badly reasoned and written report won't improve for being split into ever more sections and sub-sections.

Sawyer's Internal Auditing advises not just clear structure, but one that discourages repeated content.

> Poor organization of reported material is an impediment to clarity. Reports should flow easily from beginning to end. They should not contain closely related material in different sections. Many reports deal with complex subjects. Some report sections bear a relationship to others. To the extent possible, each audit report should be so organized that all the auditor has to say on a given subject appears in only one place in the report.[2]

You may write many different kinds of reports: audit, risk, regulatory, compliance, or board reports; project updates; briefings or memos. Yet all will have one thing in common – the readers will be human, with all the limitations we've discussed earlier. They will be tired, distracted, and under pressure from many fronts. So whatever kind of report you write, the structure should be one that leads the reader effortlessly through a coherent, compelling message.

The following structure is simple, but it's a good place to start.

Title

Executive summary

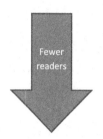

Observations or findings (possibly accompanied by recommendations)

Appendices

The report starts with a title, which may seem obvious. Yet I've been surprised how many report-writers get it wrong, being either too vague ("Payments") or too detailed ("Results of End-to-End Investigation into International Payments 1 October–31 December 2014"). Why not just say "International Payments Q4 2014"? We'll cover striking the right balance between too little and too much detail here and in the following chapters. It is often the difference between keeping and losing the reader.

The executive summary – often called "summary," "key messages," even "conclusion" or other terms – comes next. Chapter 8 will cover executive summaries, but for now, the most important thing is that it should come first. Think of a newspaper article. You'll read the headline (the title) and possibly a sub-headline (part of the title). The first paragraph – called the "lede" in US journalism – must draw the reader in. After reading that brief first paragraph, the reader should both understand the high-level message and know whether he or she wants to keep reading. The executive summary, whatever you label it, must fulfill the same role.

If you put your executive summary (or equivalent) at the end, readers will learn to skip straight to the end. If you bury the high-level message of your executive summary in unnecessary prefatory material, you will dilute it.

You'll see in the graphic above a horizontal line after "executive summary." This is because I've found that most people tend not to read beyond that point – they rightly assume that the most important information is there. Below the red line a crucial dynamic occurs, one that is essential for you to understand in order to structure your reports usefully.

Those who read beyond the executive summary will be (1) fewer and fewer as the report goes on, and (2) those who are most interested in the detail. This means that by structuring your report as above – high-level messages above the red line, increasing amounts of detail below – you will meet a variety of readers' needs.

Think about your readers, both internal and external, and you'll immediately see how different they are. They have different backgrounds, education, professional credentials and experiences, roles, priorities, and therefore reasons for reading your report.

The only way you can address those needs is by following this structure. The executive summary, as you'll see in Chapter 8, should convey a coherent message that all relevant readers – no matter how diverse – will immediately understand.

Beyond that point, your findings or observations and recommendations will feature the detail that specialist readers need and want. By including detail later in the report, you provide it while not letting it clutter the high-level executive summary, which risks distracting or boring the majority of readers. The next chapter will cover findings and observations, different approaches to writing them, and pitfalls to avoid. After all, if you don't have any individual points to communicate, you don't have anything to report.

Ara Chalabyan is President of IIA Armenia and the former chief audit executive of the Central Bank of Armenia. He is also a consultant on internal audit to the International Monetary Fund, as well as a lecturer in finance and auditing at the American University of Armenia. Previously an internal auditor, he sees reports from both sides – as a writer and as a reader. His focus is not only on clarity, with appropriate graphics (which we'll discuss shortly). It's also how to convey different information to different readers: "Concise and relevant reports with visuals are very useful. It is important to understand the users of reports: for executives, it's better to be concise; for line managers and implementers, there might be more detail and explanation." The structure above allows writers to do exactly that.

Finally, you'll see "appendices" in the suggested structure. You may already have these in your standard templates, and they can be useful for several reasons.

First, many report-writers rightly put in appendices material that others would be tempted to include in the executive summary. If you find yourself cluttering up the headline of your report with lots of background, or content from the terms of reference, you risk losing the reader. And if you lose the reader, you lose the opportunity to persuade him or her to improve the organization.

Second, appendices are the perfect place to put information that the reader – or you – may need, but not immediately. For example, if you refer to a recent piece of regulatory guidance, and one or two of your readers aren't aware of the detail, put it in the appendix. Why would you delay the majority of your readers on information few of them want?

Other readers may be inveterate trouble-makers, prompting you to write defensively. We've all come across them – the ones who want to object to your test results or conclusions, but have no factual basis for doing so. Often they simply don't want the report to convey that their department or project isn't doing well. (This goes back to communicating clearly through using active voice.) A common response from report-writers is to anticipate those objections and seek to overcome them pre-emptively by including lots of detail. "Look," the writer says. "My findings and conclusions *are* sound! To prove it, here are seventeen paragraphs of mind-numbing detail, spread over seven pages, showing how well I understand the very process I'm assessing."

This is no exaggeration – I did once read 17 paragraphs of mind-numbing detail, spread over seven pages, included solely to try to appease one or two senior trouble-makers. However, those senior trouble-makers are unlikely to change their behavior as long as you appease them. (This point takes us back to corporate or team culture.)

My experience is that stating what I've found in plain language, using the active voice and sufficient relevant supporting evidence, works. Then, if someone tries to claim that my findings are flawed because I didn't understand what I was testing, I can point to the detailed information in the appendix. It closes down the spurious objections and provides information for those who are genuinely interested. Most importantly, this approach doesn't delay or distract the majority of readers, who want to read only the relevant information as easily as possible.

A good analogy is to think of the appendix (or appendices, if you have multiple elements) as the attic of your report. You wouldn't keep skiing equipment in the entrance to your house all year round for visitors to trip over. Similarly, you shouldn't keep information that is unlikely to be immediately useful in the body of the report. The usefulness is key here – I'm not saying you should throw away your skiing equipment if you occasionally ski, or references to detailed regulatory updates if a few readers may want to see them. It's all about storing them where they're accessible, but not in everyone's way.

Speaking of regulatory updates, it's worth covering how such mandated requirements may affect the structure of your report. It's a good idea, as stated in Chapter 5, to consider regulatory requirements in the planning stage, so that you factor them in early. Look into government guidance on reporting, not just to comply with legislation, but also to make reports more effective.

The UK Department for Business, Energy and Industrial Strategy recently published a white paper (a policy document setting out proposals for future legislation) entitled "Restoring trust in audit and corporate governance: Consultation on the government's proposals." Although it is primarily about external audit, there are sections entitled "Influencing the corporate reporting framework: Promoting brevity and comprehensibility in accounts and annual reports" and "Auditor reporting."[3] These sections address "the length and volume of company reporting, raising questions about its ability to deliver meaningful information to stakeholders" and state that "auditors should provide users with more meaningful and useful opinions and information."

The point to retain here is that the government paper is not prescribing a particular structure and order that would achieve these ends. Chapter 2 referred to the Deloitte study on Irish financial services organizations' regulatory reports. It demonstrated that these reports varied wildly in approach, structure, clarity, and even the degree to which the organizations answered the regulators' questions.

Creating templates from scratch is in many ways easier than revising existing templates. One assurance professional who has experienced the latter is Vincent Chardot, Managing Director, Strategic Planning and Risk Management, Société Générale. He is introducing revised templates in the internal audit function, after a lengthy period looking at the function's reporting objectives, the readers' needs, and the best options.

Internal auditors' (and others') tendency to include too much information, for all readers and all circumstances, makes concise reporting difficult. As Chardot says,

> Auditors have to carry a large number of investigations, and they very often feel the need to report about their work, regardless of the relevance of their findings versus the overall risk profile of the activity. Another big tendency that we saw in our project to simplify audit reports is that auditors tend to forget who their main reader is and try to address a multiplicity of potential readers in their reports. This often results in the same information being repeated in the report with various levels of detail, or in multiplying layers of unnecessary contextual information before the main message is addressed.

This wasn't the function's first attempt to reduce page count and increase quality. "We actually had previous efforts of simplification that failed to reach their objective," Chardot says. What has led to success this time were three crucial steps.

> First, benchmark practices with competition. Another key element is to get senior management support, and we received for example direct feedback from our CEO and the Board, in video messages and taped interviews that we broadcasted to the internal audit manager community. Also, the auditors must be closely associated to the initiative, and the objectives clearly explained: we have actually seen through an internal survey that they were the first ones asking for change.

The first three chapters of this book focus on culture. Unsurprisingly, Chardot and his colleagues discovered that without considering culture, no project will succeed. "Changing culture is key," he rightly says, "and this must be seen as a long-term effort, that associates continuous training and communication efforts towards the readers of our reports."

The solution is not to create a one-size-fits-all approach to your reports – different audiences will have different needs. Nor is the solution to try to create a structure that encompasses every single possible reader's need into a single document. The section on templates below will discuss the unhappy outcomes, both for writers and readers, of that approach.

As with thinking deeply about your message and its recipients, and communicating in plain language, report structure must be deliberate and tailored to readers. Depending on your area of expertise, your professional body may advise on what you must include, and in which part of the report it belongs. Keep in mind that just because you must include information doesn't always mean you must include it in the executive summary – you may want to store it in the attic. If a regulator or other party requires you to include non-headline material in the executive summary, Chapter 8 will show how you can comply succinctly.

Below are a few of the many resources that may be useful to you, depending on your professional qualifications and specialist areas. In many cases, all you need to do is type a keyword such as "report" or "template" into the search function to receive a surprising list of posts, blogs, articles, and documents.

- The Institute of Risk Management (www.theirm.org/) has resources for (UK) listed companies whose reporting must comply with the Financial Reporting Council's UK Corporate Governance Code
- The International Compliance Association (www.int-comp.org/about-ica/), especially its Insight page
- The Information Systems Audit and Control Association (www.isaca.org/) has several brief posts about writing reports, especially for non-IT audiences; you can follow links to full articles for more detail
- The Association of Chartered Certified Accountants (Global) www.accaglobal.com/gb/en.html features information and resources for internal auditors as well as accountants

Don't forget to consult national or regional chapters of your professional association, so that you see both global standards and local guidance. For example, as a certified internal auditor, I belong to both the Global Institute of Internal Auditors and the Chartered Institute of Internal Auditors (UK). I also have access (as a trainer) to the French chapter (IFACI). Looking at what is available on all the sites gives me a better idea of both worldwide and nation-specific standards. The latter is particularly relevant when reporting on topics that vary according to jurisdiction: financial regulation, data protection, taxation, and so on.

Komitas Stepanyan works not only in a specialist field, but in a regulated environment. He always considers relevant standards and how they can help him better communicate to non-specialists, but also leaves room for individual judgment and creativity:

> If we are talking about reporting something about IT or information/cyber-security, there are standards like ISO27001, which has clear structure. If there is no standard requirement, such as IFRS in accounting or 27000-series in information security, then report-writing is like an art for me – there should not be any limitations to creative solutions and imagination.

Stepanyan is not the only cybersecurity expert mindful of how reports can fail to convey the very information their readers need. Jeffrey W. Brown, CISO for the State of Connecticut and cybersecurity advisor to boards, shares Stepanyan's focus on putting the reader first and telling a story.

This is a topic that a lot of people get wrong. Security professionals share whatever information they can find or whatever is easy to compile. They provide way too much data and they don't provide enough business context around what the data means. This is a case where you need to start with your audience in mind and think about what they need or want to know from you. Don't bury important information in the details or give them too much information that isn't relevant. I personally like to start with a narrative instead of raw metrics and reporting. What is the story you're trying to convey and then make sure the information you're providing supports your narrative.

During the 2020 pandemic, many teams sought to streamline their written communications, to take into account new working conditions. As people communicated more by videoconference, many built stronger relationships and shared more information with distant colleagues. One of the results was the feeling that teams could reduce the amount and nature of formal reporting. Nothing had prevented them from adopting a more concise and reader-friendly approach before – but often it takes a crisis to make progress.[4]

Several internal audit teams are now using "report on a page" or "one-page reports." We can all see the appeal – who wouldn't want to read one page instead of 10, 20, or more? Even if these one-pagers come with lengthy appendices, the principle of summarizing onto a page is excellent. However, in moving from one extreme to what some may see as another, how can we make sure one-pagers are useful?

First, as with any piece of writing, we must consider the purpose. Do we even need to issue a report, no matter how short? After all, the Institute of Internal Auditors' Global Internal Audit Standards don't say you have to communicate your findings in a report – only that you have to communicate them. As long as you include the objective, scope, and results (all of which you can summarize in a few sentences), you're adhering to the Standards.

Let's consider some other situations – and, as always, these are not internal audit-specific. You could as easily encounter them as a risk, compliance, cybersecurity, or even project management professional.

Imagine you've reviewed something in a team and found everything going well. Projects and tasks are on time and within budget while achieving objectives. The team has enough people and morale seems good. The managers use a good combination of supervision, training, and regular communications to keep everyone's skills and knowledge up to date.

In that case, why do you need to fill a page? Maybe an email will suffice. You can mention that you discovered no problems within the scope of your review and are happy to discuss details with anyone interested. Marisa Melliou thinks that one-pagers are what senior people want to see. Above all, she advises against "submitting multiple page audit reports to justify that your audit work was good. My experience so far has shown that less is more. A one-page executive summary with the most critical issues is sufficient to attract business leaders' attention."

At the other extreme, the review has produced a complicated or even contradictory picture. In this case, you will need to be extremely careful in your language – not to avoid giving offense, but to make sure every word is conveying an exact meaning. In this way, you can give more detail and context where needed, without getting bogged down in trivia, or falling back on corporate waffle. If you need to go to two pages to achieve this, is it the end of the world?

As always, keep the readers in mind, along with their possibly multiple expectations. John Chesshire appeals for writers to use the format that works best. Sometimes that will be a one-pager, sometimes more.

> Our audiences often have contradictory needs and wants. I think we can all agree that concise reporting is important, but we can go too concise for some of our stakeholders. As an audit committee chair, I want reassurance that the (internal audit, risk management or other) work has been done to a high standard, that any material issues have been identified and solutions agreed, and that where there is good practice, that this has been recognized and shared widely. You can't always get all of this from a one-pager, or an elevator-pitch – but neither do I want to plough through 30 pages of low-level dirge to find it!

LAYOUT

As mentioned at the start, this is about the "look and feel" of a report. It should draw the reader's eye, make him or her feel comfortable and bring the most important information to the fore. All too often, though, the layout becomes more about adhering to corporate style in terms of colors, fonts, and logos, rather than about readability.

This isn't solely about people's aesthetic preferences. All these elements – colors, fonts and font sizes, even margins and spacing – can make the written word more or less accessible to people. When we speak about accessibility, we often think of taking people's disabilities or special needs into account – visual impairment or dyslexia, for instance. However, making things accessible for an intended group often results in unintended benefits for everyone. Consider wheelchair ramps outside buildings; like accessible documents, these help organizations comply with anti-discrimination legislation in different countries. Yet people who don't use wheelchairs can also benefit from ramps: parents pushing children in strollers, for instance, or someone able-bodied but carrying a bulky load that may make climbing stairs risky.

Similarly, if you present your writing in clear, uncluttered documents, all your readers will benefit. Even a reader without dyslexia and blessed with perfect eyesight will not appreciate having to peer at a page of dense, unbroken text in a small, elaborate font. I mentioned earlier having seen board and committee reports hundreds of pages long, often in tiny text. It wasn't only the length that prevented scrutiny – it was also the sheer headache-inducing effort of reading six-point font.

Think of your own reaction to such pages – do you anticipate reading the content with joy or dread? Contrast this with your reaction to more reader-friendly pages, featuring lots of white space, and larger and less elaborate fonts.

There are so many resources intended to make documents more accessible, but from which we can all benefit. The endnotes of Chapter 5 include resources for people with dyslexia. However, advice on producing accessible documents will broaden access for all. There are many sources of information, and unsurprisingly, colleges and universities are often at the forefront of providing evidence-based advice and tools. Edinburgh University Information Services' "Creating accessible materials" web page is excellent, as is the University of Washington's "Creating accessible documents."[5] The Plain Language Association International runs events about plain language as a means to promoting civil rights, and Martin Cutts devotes a chapter of *The Oxford Guide to Plain English* to making language accessible to various groups.[6] Whoever the intended audience for accessible documents, everyone benefits.

The most common pieces of advice from these different resources include using sans-serif fonts rather than serif fonts, as the former can be simpler and easier to read. Serif fonts can be reader-friendly, as long as they are not too elaborate. The next time you draft a document in Word, compare the text in different fonts – for instance, Times New Roman (a serif font) with Arial (sans-serif). You will see how sans-serif produces a cleaner, plainer line, especially on screen.

Font size matters, too. Depending on the type of font, most documents should be in at least ten-point if not larger. People can always use the zoom function on computer screens to increase the size of the page in front of them, as they might use a magnifying glass with paper documents. However, few hard-copy documents handed out at board meetings come with magnifying glasses!

Finally, contrast is important. The easiest way to achieve this is to use black text on white background. Some organizations, though, come into conflict with their communications and marketing departments over this. If the marketing people have decided that the corporate style is a white font on a red background, this will cause eyestrain to many readers. The worst example I ever saw was a report in pale brown, elaborate, almost copperplate script on a lavender background. When I asked the team why they'd done this, the answer was that they felt their reports weren't eye-catching enough. The result was possibly aesthetically eye-catching – but literally unreadable.

Report layout doesn't rely only on text – sometimes, words aren't the way forward. Visuals such as photographs, charts, graphs, and line illustrations can bring a topic to life and vary how you deliver information to readers. Keep in mind that the colors you use in your visuals may or may not help readers. Those with color blindness can have trouble distinguishing red and green, making "traffic-light" or "RAG" (red-amber-green) status reporting charts inaccessible. One solution is to make sure you include the words "red," "amber," or "green" on or around the colored sections. Another solution could be to think of a fresher way to convey the message, possibly with images. After all, the now-classic traffic-light approach, while appearing to be the easy option, can provoke strong emotional responses in readers. You may end up spending more time arguing about the color than discussing substance.

Interactive reporting can even allow readers to click on charts and links in the electronic version of a report to see further detail. This is an excellent way to keep the body of the report uncluttered. Vertical lists using bullet points or numbers are also much more useful than a rambling list over several lines of a single sentence. The list approach saves the reader the trouble of working out the different elements of the sentence and their relationships; it presents a clear number of items, distinct from each other.

Varying how you present your information, while keeping it clear and uncluttered, will appeal to the widest relevant range of readers. Komitas Stepanyan points out that sometimes

> you need to communicate something without knowing who your audience is. Different people may receive, accept, and analyze the same information in different ways. For example, I may prefer long, detailed text, while my colleague may prefer more visuals, and graphs instead of numbers, in text with detailed explanations.

Speaking specifically of reports, he shares his experience of creating two documents to meet readers' needs:

> While working in the Internal Audit department at the Central Bank of Armenia, we were doing a presentation for each audit engagement. It was not easy to write reports in Word, and then create a PowerPoint report. Sometimes it may have been seen as time consuming. However, if this approach helps you fulfill your objective, then any effort, any way of communication is useful.

His colleague Ara Chalabyan shared the same experience:

When I was in an assurance function, I asked my team to be creative. We switched from MS Word reports to slides with visual content to the extent possible. There is no standard for good writing but inquisitiveness, the constant search for excellence. Do not copy and paste old templates: use creative tools, animation, transitions, highlighting, photo, and video.

Showing that this approach knows no borders, Marisa Melliou is also a fan of simplifying reports not only through writing but also through the medium. She wants report-writers to "Break down complex or highly technical concepts into relatable terms, while also turning facts and data into a compelling, colorful narrative that the business can act upon. Use PowerPoint presentations or videos – no more MS Word paper reports!"

If report-writers are to do all this – think carefully, use plain language, plan rigorously, and structure reports adapted to readers' needs and expectations – they will need to spend time and energy doing so. Can standardizing report templates help them recoup some of this time and effort, and be more efficient?

TEMPLATES

Most teams have standardized templates for reports and other regular writing. It saves time, fosters consistency, and often helps new members of staff understand what to write where. There are many excellent reasons to have standardized templates – as long as we keep in mind that they are a means to an end, not an end in themselves.

This means that templates stop being useful the minute people justify including irrelevant or repetitive information by saying the template requires it. If you have templates that encourage or compel poor communication, change the template.

Having said that, many people have complained to me that they write badly because they're forced to use a bad template. Sometimes this is true. Other times, the template is fine – the problem is what people are writing in it. It also happens that blank templates feature sound guidance in each section, stating clearly its purpose and what writers should include. However, it's common for people to read the guidance the first time they complete the report template, and never again. Instead, each time they write a new report, they simply replace the previous report's content with new content and save the new version. Any memory they have of the guidance fades as the weeks, months, and years pass, until they claim no knowledge of any guidance, whether at team or organization level. As Rachel Browne says, "We have a corporate style guide. People don't always use it or even know it exists. We use report templates which are meant to be tailored, but people see a section in the template and think they have to write something for every section." The result, unsurprisingly, is unnecessary detail and disengaged readers.

When this happens, teams often notice reports aren't having the effect they should. Rather than stripping the template back to basics, though, they compound the original problems by adding yet more unwanted content. As I write this book, one of my clients is reviewing its report template, which has grown over the years to nearly 30 pages when blank!

The template has developed organically and, as various readers have mentioned not being able to find information, the client has added more sub-sections and sub-sub-sections. This goes back to the signpost analogy – instead of making the content clear, the client is simply adding more opportunities to include and repeat unclear content. The repetition inherent in this kind of template solves no problems and creates many more. First, repetition turns readers off – they see information they've already seen and think, "I know this – there's nothing new after this. I can stop reading now." If they keep reading, they may notice inconsistencies. This client has several sections to the report in which the same information (or much of it) is presented in slightly different ways. If we imagine that the report covers topics A, B, C, D, E, and F, the client's current template does this:

- Section 1: A, B, C, and D
- Section 2: A, B, C, D, E, and F
- Section 3: B, C, F, and A
- Section 4: D, E, and sub-sections of A and C
- and so on (for 30 pages)

You can imagine how hard it is to follow, let alone infer a coherent message. I have several times had to spread the pages of the reports out across a desk and draw color-coded lines among them to track what appears from one section to the next, what disappears, and what seems to transform into something altogether different.

Few readers will analyze the reports in such depth. As we mentioned earlier, they will stick to the executive summary and trust this gives them what they need. However, you can see from my description above that the executive summary does not appear to include all the relevant information. And if your template is 30 pages before you even begin to populate it, how can you – let alone the poor reader – keep track of your content?

Even if you do, other inconsistencies will appear. If your report template requires you to refer to specific findings or observations in three or four different places, then, if you receive last-minute changes, you will have to carry those changes through the whole report. If, being human, tired, and under pressure, you overlook one of those changes, someone is bound to notice. "You said there's a 67% failure rate here, but in the executive summary it says 82%. What's going on?"

You can all imagine the stress and confusion over-engineered templates can cause. The best thing is to choose the most basic one, to which you add only if you must.

Different organizations will have different standards and expectations; communications formats and templates will also differ. One security aware-ness professional, with experience serving in the military, has compared reporting in the army and in the private sector, concluding that quality suf-fers when standards are inconsistent.

> The standard of reporting in the corporate sector is significantly lower than what I saw in the military. The military uses a standard template to produce documents across all three services and joint units. This creates a simple process to follow, and ensures issues and recommen-dations are clearly articulated and understood. Equally, senior deci-sion-makers expect the quality of written work to be high and enforce such standards. This is not the case in the corporate sector (or at least what I experienced). The diversity of personalities and approaches, with no enforced standardization – unless a report was going to senior executive level – meant the general quality of reporting was mediocre at best, and often difficult to understand. However, a positive attribute of "report writing" freedom is creativity. The lack of standardization in the corporate sector, that I saw, meant employees had the opportu-nity to express ideas in different ways. This opened up the opportunity for creativity, and did make report writing more engaging for a wider audience.

There are pros and cons to all approaches; consider what works best for your readers, in your organization. Simplicity is always a good basis, whether in language or in structure. Keep it simple and you can then easily and clearly develop where necessary. A streamlined template allows for exactly the kind of creativity several experts have called for in this chapter.

Next are three different sample templates: one I created; one from a for-mer client; and a one-pager from another client. (The client documents appear anonymously and with permission.)

All three are from and reflect an internal-audit context, but you can adapt them to your own needs, simplifying the language where possible.

Report title and reference number (if applicable)
Month DD, YYYY

Responsible executive's name and full title

Executive summary		
Audit rating Rating goes here **Prior Audit Rating (Month, Year)** Rating goes here (remove if NA)	**Findings by risk level** High: 0 Medium: 0 Low: **0**	**Risk assessment** Inherent: H/M/L Residual: H/M/L

Business overview (optional)
Provide a brief (one to two sentence) overview of the audited entity: high-level description and size metrics. If you feel readers need a reminder of objectives, scope or methodology, consider including the terms of reference as an appendix.

Conclusions
One or two short paragraphs. The first paragraph should answer the question, 'What do we think of how this area manages its risks?'. Conclusions should provide rationale and justification for the selected report rating and should not simply summarize individual findings. Make clear whether the client has accepted our conclusions fully, partly or not at all.

Summary of high- and medium-risk

	Risk Rating	*Subject*	*Action owner and due date*
1	H/M	Observation	Full Name Full Date
2	H/M	Observation	Full Name Full Date

Sample template A, page 1.

Detailed findings (in order of severity – high, medium and low)

Finding 1 title: This should give enough information so as to be meaningful if extracted to appear in MI. 'People' or 'Cash handling' is not enough – 'Unclear recruitment process' or 'Inconsistent cash-handling procedures' is more useful.

Observation: This should answer the question, 'Who is not doing what?' This will almost never be an individual – rather, a team, a department, or senior management. If a control is not in place, then you can word it in terms of 'X has not put in place Y.'

Risk: This should answer the question, 'Why do we care?', and should articulate a clear financial, reputational, regulatory or safety risk. It should not list further failed controls.

Root cause: This should answer the question, 'Why is the control inadequate or ineffective?' It must not repeat or rephrase the observation.

Additional information (optional): This section can be useful for further detail – test results making clear the materiality, or regulatory guidance. Its purpose would be to keep such detail out of the observation. If you need to include extensive detail – pages of MI, for instance – that can go into an appendix.

	Management action plan	Action owner and due date
1.01	Each element of the action plan must address root cause.	Name Unit Due Date
1.02	Each element of the action plan must address root cause.	Name Unit Due Date

Sample template A, page 2.

Distribution and Audit Team

To:

Cc:

Audit team:

Sample template A, page 5.

Appendices

This is for additional information that does not *need* to be in the body of the report. Items you can usefully put here include:

- Terms of reference, or elements of the ToR (object ives, scope, timescales, methodology etc.)
- Regulatory guidance
- Benchmarking documents
- Policy documents
- Process maps
- Detailed test results
- Lists of people consulted or interviewed

Sample template A, page 6.

ORGANISATION
INTERNAL AUDIT UNIT
GOVERNMENT DEPARTMENT
BUSINESS UNIT

CONTENTS

VERSION CONTROL

Version	Document Description
X.X	Pre-draft, draft or final report

Ref: Year Audit00X
Date: Day Month Year
Audit Team:
Name, Title

MANAGEMENT

Sample template B, cover sheet.

PURPOSE OF REVIEW

1. Formatting throughout is single-spaced, with 6pt leading (spacing) before and after each paragraph.

2. Heading above (Purpose of Review) is all caps, 18pt Calibri 18pt, blue.

3. Text of body is 11 pt. Calibri, black.

4. Justification is full throughout. Left margin is 0, with hanging indent at 1, for all sections (including sub-headings and sub-sub-headings).

5. Have a space between each major section (Purpose of Review and I, II, III etc.) – as below.

I. EXECUTIVE SUMMARY AND ASSURANCE STATEMENT

6. Text of body is 11 pt. Calibri, black.

7. In light of the above, we are able to give Full Assurance/Moderate Assurance/Partial Assurance/**No Assurance**.

   We include an Assurance Statement within each of our internal audit reports. The Assurance Statement is used to highlight our overall view for stakeholders of the adequacy and effectivness of the risk management and internal control system in respect of the team, system, activity and/or process under review. Assurance is based on the following criteria:

   Full Assurance: Risk management and internal control activity is sufficient to secure the consistent and effective achievement of the objectives of the team, system, activity and/or process. The impact of any non-compliance with key controls is low.

   Moderate Assurance: Strengths in risk management and internal control outweight weaknesses. Although there is a need for improvement in specific areas, the team, system, activity and/or process generally operate/s effectively.

   Partial Assurance: There is a risk to the achievement of the objectives of the team, system, activity and/or process. Some of the key controls are either missing or not operating effectively.

   No Assurance: Control is insufficient to secure the consistent achievement of the objectives of the team, system, activity and/or process. Some of the key controls are either missing or not operating effectively.

II. KEY FINDINGS AND RECOMMENDATIONS

IIIA. SUB-HEADING IN 14 PT CALIBRI, BLUE, ALL CAPS

Sub-sub-heading (only if absolutely necessary!) is 14 pt Calibri, blue, not all caps

8. Text of body is 11 pt. Calibri, black.

9. *Recommendation – Desirable:* (or *Recommendation – Highly Desirable* or ***Recommendation – Critical***): XXX.

   Benefit: XXX

Sample template B, page 1.

Recommendations and management actions are included in our reports to highlight and address areas of particular internal audit concern. They are principally based on our assessment of the:

- risks that the team, system, activity and/or process may fail to deliver its objectives;
- likelihood of the risk materialising; and
- impact on the business of something going wrong.

Recommendations and management actions have been categorised according to their level of importance:

Critical	To address a weakness (in the team, system, activity and/or process) that would materially affect the business, or would appear in the HIA's annual report.
Highly Desirable	To address an important weakness in the team, system, activity and/or process, but one that does not materially affect on the business.
Desirable	To address a weakness so as to improve risk management or value for money in the team, system, activity and/or process. Without this improvement, however, the business will continue to achieve its objectives but with limited efficiency.

IIIB. SUB-HEADING IN 14 PT CALIBRI, BLUE, ALL CAPS

10.   Text of body is 11 pt. Calibri, black.

11.   *Recommendation – Desirable:* (or *Recommendation – Highly Desirable* or **Recommendation – Critical**): XXX.

   Benefit: XXX

IIIC. SUB-HEADING IN 14 PT CALIBRI, BLUE, ALL CAPS

12.   Text of body is 11 pt. Calibri, black.

13.   *Recommendation – Desirable:* (or *Recommendation – Highly Desirable* or **Recommendation – Critical**): XXX.

   Benefit: XXX

Sample template B, page 2.

APPENDIX A: AGREED IMPLEMENTATION PLAN

The following entries represent the agreed implementation plan for the recommendations and management action identified during the course of this internal audit. This implementation plan has been completed by the XXX, and agreed with XXX.

Paragraph XX	*Recommendation – Desirable* or *Recommendation – Highly Desirable* or *Recommendation – Critical*	
Copy and paste your recommendation from the relevant paragraph. As before, the spacing is 6pt before and after each paragraph, and single spacing for each line.		
Within the Table Properties, all content is fully justified.		
Accept	Target Implementation Date:	Owner:
Client Response:		

Sample template B, page 3.

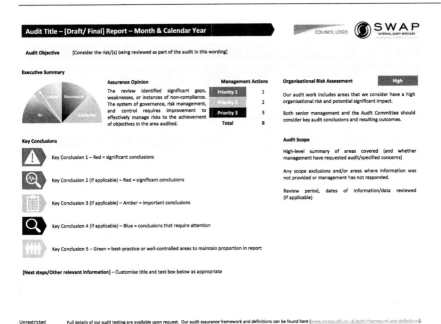

Source: SWAP Internal Audit Services

Next are two further examples of one-pagers. The first is for an audit report, the second for a report to the audit committee. You could adapt this approach to your internal and external readers, including steering committees.

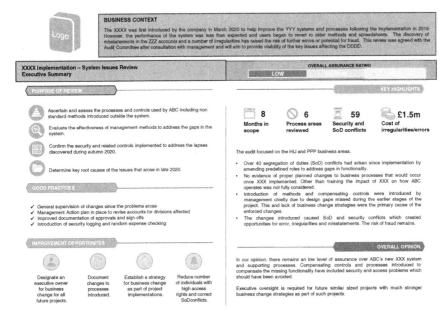

Source: Malcolm Zack, Duyen Nguyen, and Scott Petersen. Designed for an organization in the testing and certification sector. The information provided is based on a hypothetical scenario and has no relationship with the company concerned.

ABC Process

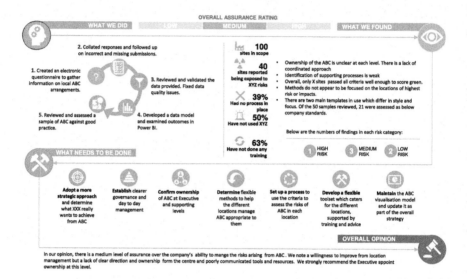

Source: Malcolm Zack, Duyen Nguyen, and Scott Petersen. Designed for an organization in the testing and certification sector. The information provided is based on a hypothetical scenario and has no relationship with the company concerned.

If you already have a template and are unsure how to make it more effective, try the following. Create a "focus group" of six to ten regular readers of your reports. These people may be board or committee members, senior managers, or even operational experts responsible for putting recommendations in place. Choose people who will give you honest, unbiased, measured, and informed responses. Ask them the following:

- When you read our reports, what do you always turn to first?
- What do you always skip?
- And what do you wish we could include that we currently don't?

The answers may surprise you, but they will also help you produce a more effective template. Your respondents' answers to the first question will show you where you have perhaps too much "filler" at the beginning of the report. If your executive summary starts with sub-headings such as "introduction," "background," and "context," most readers will quickly learn where the real substance is. They will turn immediately to page three, five, or even seven, if that's where your actual executive summary lies.

If this is the case, your task is simple: move whatever they read first to the front. If the "filler" that currently delays readers is for any reason truly necessary, think where else you could put it. Does it belong in an appendix – in the attic?

The answers to the second question should also lead to an obvious change in the template. If your most thoughtful and responsible readers always skip something, you can probably delete it. If you're including it for legal or regulatory reasons, consider storing it in the report's attic.

The third question prompts some of the most amusing responses – amusing, because often people wish you'd include something that's already there! However, that is useful information. If you are including information, but readers don't realize it, one of two things is happening. Your readers aren't reading far enough into the report, or they're reading the whole report, but don't realize you've included what they need. In either case, you must ask why. Likely reasons include wordy, convoluted writing, and misleading or repetitive report structures. Identify what is preventing your readers from finding the information they want, and correct it.

The example below comes from a company whom I've advised about creating more reader-friendly templates. You'll see that the order reflects the structure shown at the beginning of this chapter. What's more, it's the result of exactly the type of readers' survey I've just described.

Email Subject Line: Audit Report Name - Internal Audit Report 202X-01

Month XX, 202X

RATING
For information on rating definitions, please see *hyperlink*.

Executive Summary:
XXXX (the Company) has appropriately managed its resources through a XX% decline in [XXXX metric]. Areas that are operating effectively are XXXX. Minor improvement is needed in the areas of XXXX.

Observations: For full details, please see *hyperlink*.
#1 [Summary Description of Issue]
High Impact/Low Likelihood

#2 [Summary Description of Issue]
High Impact/Low Likelihood

#3 [Summary Description of Issue]
High Impact/Low Likelihood

In addition to the above report observations, there were X Minor Observations and X Process Improvements. For additional details, please see *hyperlink*.

Background:
The Company is part of the XXXX Business and is a leading producer of XXX systems, XXXX, and XXXX systems for domestic and international markets. Total 202X sales were $XXM (domestic $XXM, international $XXM). For full details, please see *hyperlink*

Scope:
We audited XXXXXX. For full details, please see *hyperlink*

The uniquely helpful thing about this format is that it's all within the body of an email, with hyperlinks to detailed findings. Readers don't have to open any attachments – simply scroll and, if they wish, click.

A final image comes again from Zack, Nguyen, and Petersen. It shows how interactive reporting – in this case, using business intelligence (BI) tools, can allow readers to click on a particular element and see a detailed window pop up.

Interactive Internal Audit Report (Pilot)

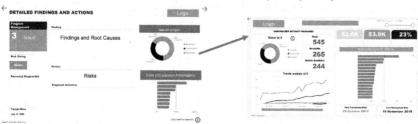

Interactive Power BI report – Extract.

Detailed finding section on left has normal audit information. Graphics on right side of that page provide initial information that the user can click or drill into. Opens up supporting dashboard which supports findings, root cause and risk points

Source: Malcolm Zack, Duyen Nguyen, and Scott Petersen. Designed for an organization in the testing and certification sector. The information provided is based on a hypothetical scenario and has no relationship with the company concerned.

Providing what people need in one page – or, even better, in one email – gives them the essential information at the start and, crucially, choice. Rather than forcing everyone to work through page after page, hunting for something relevant, this format allows readers to select as much or as little detail as they need.

Video reports have become more popular in recent years. Larry Herzog Butler, Global Head of Internal Audit at Delivery Hero, based in Berlin, introduced video reporting during Covid-19 lockdowns. They now produce brief – two- to three-minute – videos at seven- to ten-day intervals throughout audit engagements. These videos update viewers on the progress of the engagement, good

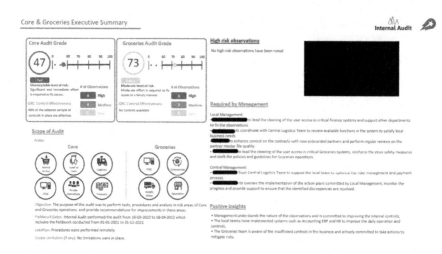

Source: Delivery Hero SE

news, and areas of concern. The response, says Butler, has been extremely positive, and the team's internal audit report includes a link to the video.

Another audit leader who uses videos – currently in conjunction with full written reports – is Michele Variale, Chief Audit and Risk Officer at Telepass. In an edition of The Audit Podcast, hosted by Trent Russell, he discussed his team's experience of communicating results by video.

> It's a good way to also test how strong you are on your beliefs about the results of the audit, how powerful you are on your storytelling, which is not a list of issues, it's a business discussion with the CEO on the impact of a process or any control breakdowns on the achievement of the objectives of the organization. We need to step up, and by using a video, we are forced to step up.[7]

Initially, Variale states, people were a bit skeptical – as they had been with other changes such as "agile" auditing, use of robotics, and combined assurance. However,

> then the CEO and Chairman saw huge benefits for the simple fact that it's freeing up time. They can view the video in a taxi or queuing for a flight, immediately get a flavour, understand the level of attention needed for the audit report.

The videos always end with an invitation to contact the team with any questions – so rather than a written report acting as the final word, the video is an invitation to a conversation. His team is now seeing more response to videos than to previous methods of communicating results.

As Trent Russell says, different media can make your readers' lives easier and convey understanding and respect for them. "In being able to send a video, you've made a point about respecting that person's time. It also frees you up from having to schedule a meeting," he says to Variale in the podcast. And, he wonders, how often will people now leave a meeting thinking, "[t]hat meeting should have been a video"?

Liz Sandwith is an internal audit and risk consultant. Her distinguished career includes leading internal audit at Bupa, Channel 5 (UK) and the UK Electoral Commission, as well as advising a number of local authorities and the Information Commissioner's Office on internal audit and risk management issues. She has seen every format and fashion of reports come and go.

> Based on 33 years in internal audit, I would suggest that written communication is often a challenge for internal auditors. Reports often vary in length from 40+ pages to the new, more concise approach, with reports being five pages or fewer. However, whatever the length, the challenge is still present. Sometimes more words confuse and muddle the message – as can reports that are too short. A CEO once said to me

that if you cannot convey your message or story in a page, then you probably don't understand the message.

Improving your report templates using your readers' views obviously will help meet their needs. But it also benefits the team producing the report. Whenever clients tell me they have recently updated their report templates, I notice that the team members seem resentful and exhausted. This is often because certain people will have worked hard to create the existing template. When the team as a whole then discusses how to change it, it's natural for people to feel protective and defensive of their work. This then turns what should be a positive exercise – improving what the team produces for readers – into a closed and often heated debate within the team.

Looking outward changes the dynamic, and de-personalizes the discussion. Instead of someone saying, "We need to change Sara's template," it becomes, "Eight of our ten focus-group readers turn immediately to page five. The most useful information for them is there, so can we move it to page one?" As a result, the process of improving the template achieves its aim without damaging individual or team morale.

Some teams tend to avoid the debate about templates altogether by leaving it to software. Program management and writing-specific software packages are available. Both collate the information available and produce a draft report in a pre-defined format. This can be useful, provided you know exactly what purpose the report serves, and which structure and layout will best meet that purpose.

Some people use "report" to mean any document showing the results of computer-assisted research, including data analytics products.[8] Others will mean reports as we are discussing them here – reports that give a high-level message supported by discrete findings or observations. If your team uses or is considering using program management software to record your field-work and test results, and produce a concluding report, make sure you can tailor the report template to your and your readers' needs. If the standard report produced by the software has too many fields and subfields, you risk repetition, error, and losing the reader.

More important, remember the classic computing acronym, GIGO: garbage in, garbage out. However good the software, if you populate fields with poorly reasoned, poorly written text, don't be surprised if the resulting report is unreadable.

Remember always – reports are a means to an end, not an activity in themselves. As we saw earlier, at least one professional body (the Institute of Internal Auditors) doesn't even mandate a written report.

In the first edition, I referred repeatedly to reports prompting action. On reflection – prompted by Mike Jacka's excellent insights on the IIA Global blog – I realized that this doesn't reflect practice; nor should it. If we are communicating throughout an audit engagement, risk or compliance review, or information security check – then what is in the report should come as no surprise to anyone, least of all the readers.

As Jacka says, "If you are still using the report to prompt the reader into taking action, you have already lost the battle."[9]

So, what do reports do, if not prompt action? According to Jacka, "The report is a reminder to everyone of what has occurred — a record of agreement." This should be the case in audit, risk, compliance, and information security. However, certain reports may indeed exist to prompt action – in the case of a board paper asking members to make a decision, for instance. An even more compelling example is that of the Intelligence Community Assessment (ICA) produced by Central Intelligence Agency (CIA), National Security Agency (NSA), and Federal Bureau of Investigation (FBI) in January 2017 on Russian interference in the 2016 U.S. presidential election. The Key Judgments section of the report concludes, "We assess Moscow will apply lessons learned from its Putin-ordered campaign aimed at the US presidential election to future influence efforts worldwide, including against US allies and their election processes."[10] Although the ICA is not in a position to dictate to the government of the day what action to take, it's clear it wishes its report to prompt action, to safeguard the integrity of future elections.

Given how many organizations rely on reports, though, the format in which they're issued matters. Structure, layout, and templates can all help or hinder understanding. If they're not aiding readers and leading to improvements, you must discover and address the reason. No matter how established standard templates or reporting habits are, if they're not working, the organization isn't working.

Think of where you can seek advice on how to make the "look and feel" of your reports more appealing and therefore more effective. We've mentioned professional bodies that provide guidance to members, as well as internal focus groups of readers. Another often underused source is quality assessment (QA) teams. You may have one within your function, or there may be an organization-wide team focusing solely on quality. What good and bad practices do they see in the organization and elsewhere? QA teams keep abreast of industry standards and peer practices, including written communications.

You may discover that pressure is affecting the quality of structure, layout, and templates. If people are so busy issuing reports that they don't have time to see if the medium is effective, then they are wasting time and effort. Reports delivered on time but in a reader-repellent format don't help anyone. As Komitas Stepanyan says, "I think no matter where you work – private or public sector – effective and timely communication is a crucial component and good indicator of how mature the organization/institution is."

Effective comes first, although timely is important. Keeping your structure, layout, and templates simple will help you deliver better reports on time, meeting readers' and the organization's needs.

ACTIVITY

Imagine you've recently joined a team, and a colleague has shared an outline of the report template. Can you identify at least six elements likely to distract or confuse readers? You don't need any specialist knowledge or further reading for this exercise – trust your judgment and common sense. (Suggested answers after the chapter summary.)

    I. Title: IT review
    II. Executive summary
        A. Introduction: an overview of the topic
        B. Background: background to the topic
        C. Context: wider organizational context, including scope of review
        D. Key summary: summary of review
        E. Key messages: summary of key messages from review
        F. Conclusions: conclusions of review
    III. Summary for committees
    IV. Recommendations
        A. Improvement A
        B. Improvement B
        C. Improvement C
        D. Action plan agreed by management
    V. Appendix: detailed findings
    VI. Appendix: test results
    VII. Next steps

SUMMARY

- Well-reasoned and well-written reports need reader-friendly formats. If your reader can't read your report, your efforts will have been in vain.
- Thinking about how you structure your reports – continuing from the planning stage – will make them more logical and coherent.
- The simplest structure (title, executive summary, findings or observations, recommendations, and appendices) is the best approach.
- Too many sub-sections and layers encourage repetition and error, and discourage readers.
- The layout must be easy on the eye. Advice from government agencies, charities, and other organizations about making documents accessible will benefit all readers.

- Templates can save time and effort, but only if they respond to readers' needs. If you find yourself populating a template with unnecessary or repetitive information because the template demands it, change the template. Seek your readers' advice on what they find most and least useful, and adapt accordingly.
- Software programs can produce reports from populated fields, but those using them must take responsibility for the template and quality of content.
- You needn't restrict yourself to traditional reporting methods - emails, presentations, even videos can all be ways of communicating your results more clearly and efficiently.

ACTIVITY – SUGGESTED ANSWERS

Elements that are likely to distract or confuse readers – did you find at least six of these?

1) The title is too vague.

2) The executive summary has too many sections.

3) What is the difference between introduction and background?

4) Context can be useful as long as it is not too lengthy. Chapter 8 will discuss this in more detail.

5) Key summary: Is this the actual executive summary?

6) Key messages: How is this different from the key summary? If it's a list of findings, this isn't the place for it.

7) Conclusions of review: Again, is this the actual executive summary?

8) Summary for committees: I have seen this only once among thousands of reports. What is its purpose? Is it because the executive summary itself doesn't summarize effectively for committees?

9) Recommendations: Why would you start with recommendations, before stating what problem they should solve? Chapter 7 will discuss this in more detail, but the lack of logic is clear.

10) Action plan agreed by management: Again, this seems illogical. If the action plan includes the recommendations, there is unnecessary repetition. If the action plan doesn't include the recommendations, then there is a disagreement between management and the report authors. This needs to be clear in the executive summary.

11) Many teams included detailed findings in an appendix. This is acceptable, but not if it leads report-writers to repeat the findings in the executive summary for fear readers won't read the findings.

12) Next steps: How is this different from the recommendations and agreed action plan? You may conclude that the template can be simpler, or that you could use another medium altogether. Once you have organized your thoughts, as discussed in previous chapters, the results should apply to any means of communication: report, presentation, video, or other.

NOTES

1 Pressfield, *Nobody*, 163.
2 747. See also Chapter 15, "Reports to Executive Management and the Board" (849–81).
3 March 2021, 82, 106, https://assets.publishing.service.gov.uk/government/uploads/system/uploads/attachment_data/file/970673/restoring-trust-in-audit-and-corporate-governance-command-paper.pdf
4 From content and attendee contributions to a Chartered Institute of Internal Auditors webinar entitled "Reporting during and after Covid-19: less is more?," hosted by Sara I. James and Liz Sandwith, July 23, 2020, October 16, 2020, and May 18, 2021.
5 Edinburgh University Information Services www.ed.ac.uk/information-services/help-consultancy/accessibility/creating-materials and University of Washington www.washington.edu/accessibility/documents/. See also advice from the UK government www.gov.uk/guidance/publishing-accessible-documents, AbilityNet (created by IBM and Microsoft) abilitynet.org.uk/factsheets/creating-accessible-documents-0, the UK charity Change www.changepeople.org/getmedia/923a6399-c13f-418c-bb29-051413f7e3a3/How-to-make-info-accessible-guide-2016-Final, and Mohawk College www.mohawkcollege.ca/about-mohawk/accessibility/tools-and-resources/creating-accessible-documents-and-alternate-formats
6 "Writing low-literacy plain English," 269–81.
7 The Audit Podcast Episode 185: Audit Reports and How They Solve Audit's Biggest Issue w/ Michele Variale www.youtube.com/watch?v=48JpbfjjVOs

8 Tommie Singleton, "IS Audit Basics: What Every IT Auditor Should Know about Computer-generated Reports," *ISACA Journal Archives*, October 1, 2014, www.isaca.org/resources/isaca-journal/past-issues/2014/is-audit-basics-what-every-it-auditor-should-know-about-computer-generated-reports

9 Mike Jacka, "Mind of Jacka: Prompt Versus Remind," IIA Global blog, March 9, 2023 https://internalauditor.theiia.org/en/voices/2023/mind-of-jacka-prompt-versus-remind/

10 ICA_2017_01.pdf (dni.gov), p. ii.

Part 3

Words into action

Chapter 7

Findings or observations and recommendations

We've covered the prerequisites of any effective piece of writing: culture, reflection, plain language, and rigorous planning. If you were building a house, you'd check the ground first and work closely with an architect on a good blueprint, before sinking solid foundations. Only then could you start building – let alone decorating – the house.

Now we come to building: to the information we include to document or prompt action. Thinking about your readers, both internal and external, will reveal different backgrounds, training, and experience. As Chapter 5 demonstrated, careful planning can help you better understand who is reading the report, for what purpose, and in what context.

Any report will convey findings or observations. These could be audit or risk findings, which some teams call issues. Compliance teams may refer to them as test results. IT and cybersecurity teams use a variety of terms, depending on what they think will best grab readers' attention. I've seen "findings," "issues," "observations," "key points," and even "weaknesses." The last one certainly uses plain language, which best alerts readers to problems. Terminology – like tone – can affect not only interpretation but also relationships, as Richard Chambers and Norman Marks have pointed out.[1]

In this book, we'll use the term "finding," as it can cover everything from neutral observations to serious warnings.

Your readers likely include different levels of management. Senior managers and board members may read reports simply for information or because they need to take strategic decisions and issue directions to resolve findings described in your report. Other managers – such as operational and middle management – will be responsible for correcting findings. This takes us back to the suggested order at the beginning of Chapter 6; everyone will read the executive summary, while those needing detailed information and recommendations will read further. You can also structure the findings themselves to order information from the broadest to the most detailed, as this chapter will show.

DOI: 10.1201/9781003422365-10

It often goes back to what we discussed in Chapter 1 – culture and communication. Remind yourself who your readers are, where they come from, and what they expect. You may find different reactions to findings depending on national or regional culture. Rinus de Hooge has found that

> in countries with a recent authoritarian regime, e.g. former Soviet countries, management takes the report very seriously and asks for instructions how to implement individual actions on findings. It is seen as a directive by audit. In countries where there is no history of authoritarianism, discussion is more open and disagreement about findings and ratings are accepted. In the Netherlands it is almost a sport to disagree first with audit and then turn around. It is rare that the recipient lifts the report above the individual findings, and tries to discuss the root cause of the situation and tries to fix that – and not the individual findings. I have seen this happening in Scandinavia where the finding itself was not that interesting, but the cause was discussed.

However interested your readers are in different elements of findings, those elements must all be present. But what are they?

There are different approaches to writing what I've called findings, depending on your function and audience. I am in the habit of applying what some may call an internal-audit-specific approach, but I've found it applies to various settings, including project management. One client, formerly an internal auditor but now in organizational security, always asks me to use this approach when training her teams, no matter their function. It works, she says, because it makes people focus on the non-negotiable elements of a finding.

The first element is an observation. This is whatever is happening (or not) that is worth commenting on. The second element is the risk – what is the tangible harm that could come to the organization from this situation? The third element is the root cause: how did this situation arise?

Some internal audit functions use an approach called the five Cs. In this approach – which is not and never has been required by the Institute of Internal Auditors – the finding includes criteria (or criterion, if singular), condition, consequence, cause, and corrective action.

The five Cs list is pleasingly alliterative, but how useful is it? Many teams use it in their reports, and their fieldwork documentation will certainly need to include at least the first four Cs. However, I believe in simplifying life, so focus on three questions – the three Qs.

- Who is not doing what? (observation)
- So what? (risk)
- How did this happen? (root cause)

This approach strips findings back to three elements without which you simply *cannot* have a finding. Including criterion or criteria all too often leads to lengthy process descriptions, before eventually allowing the reader to learn "who is not doing what." (Remember the 17 paragraphs of mind-numbing detail I mentioned in Chapter 6? It was a process description.) Put yourself in your readers' shoes – why would you want to read so much detail, almost all of it unnecessary to understanding the observation?

The fifth C – corrective action – also doesn't feature in my list of questions. This is because not all reports include recommendations. Furthermore, management action plans require senior managers to create them. It's rather unfair to require the fifth C when many teams will have little or no control over it.

This does leave us, though, with the spine of any sound finding: the three Qs.

WHO IS NOT DOING WHAT?

This is a simple way to communicate the problem. Someone – usually a team, area, or department – is not doing what they should. It could be they are not adhering to controls, following instructions, or putting necessary processes in place. This also includes doing things they shouldn't, which may seem the opposite of not doing what they should.

However, the latter case is almost always one of violating existing standards, norms, or controls. So if you say, "Staff members allow unauthorized people into the building," it's clear to any rational reader that this is a case of not following procedures (implied in the word "unauthorized"). The same is true with a sentence such as, "The team processes payments over $5,000.00 without seeking a senior manager's approval." Again, it's clear from the sentence that a senior manager should approve payments over this amount.

A finding can also communicate "What is not in place?" or "What is not functioning as it should?" However, someone – a team, department, or senior management – is likely responsible for putting something in place, or making sure something functions as it should. This is why I use "Who is not doing what?" as a way of forcing report-writers to understand who is responsible for tasks, systems, or processes. After all, it's too easy to blame a system for not being reliable – but someone is in charge of monitoring and maintaining that system. Similarly, instead of saying, "There is no IT security governance framework in place," why not find out which team or committee was responsible for the task? One finding I read stated bluntly that the CISO, who had been in post for over six months, still had not established an IT security framework. It was a rare instance of referring to an individual's role, rather than an area, team, or department. But it communicated the problem clearly and concisely.

You will usually provide more information than that in a finding, but the single-sentence approach is a necessary start. It tests your information and understanding: if you cannot summarize in one sentence "who is not doing what," then you need to reexamine your thinking. If you cannot articulate "who is not doing what" in a way that makes it clear what people should be doing, revisit your wording. In either case, the active voice will help you.

If you think back to Chapter 4, we looked at how the active voice forces us to confront any gaps or assumptions in our research. If all of our notes state, "Unauthorized people are allowed into the building," or "Payments over $5,000.00 are processed without a senior manager's approval," it's passive, and easy to overlook who is responsible. This isn't for the purpose of blaming people – rather, it's to see clearly where exactly different processes sit and how they work. Without this knowledge, any findings are like houses built on shifting sands.

Overusing the passive active voice can also, as we saw in Chapter 4, mislead readers. If you remember the example of the IT services team, the initial finding used the active voice, followed by the passive voice. ("The IT Services Team does not adhere to Control A. Controls B and C are not adhered to, either.") What the passive voice hid was the "doer" – the reader naturally assumed that the IT services team mentioned in the first sentence was responsible for the failing mentioned (passively) in the next sentence. As we discovered, this was not the case.

Make sure that during your research, investigation, or fieldwork, you write as much as possible in the active voice. This will help you see more clearly how different tasks, actions, and stages of processes work and even fit together. Then, when you start to draft a finding, you do so from a position of confidence in your material. You can always soften a message later – if necessary.

Why would it be necessary? As Marisa Melliou, quoted in Chapter 4, says, blaming individuals rarely helps, so many report-writers avoid using active voice for this reason. However, as we've seen, reports usually address a particular department, area, or even process, involving one or more teams. There's no reason not to mention which team isn't adhering to standards or following processes. After all, the readers know that these are the team or teams involved in doing so – why shy away from what should be obvious? More to the point, why risk the kind of misunderstanding that the IT Services Team example creates?

Many writers choose to risk it – where they are even conscious of doing so – because they fear conflict. Again, Chapter 4 addressed this, and the extent to which even some communications consultants will encourage short-term, high-risk responses such as using the passive voice. It's often a mistake. If we go back to the purpose of reports, it's to enable readers to do something meaningful: either taking action or thinking differently. (The latter usually enables the former.) If you obscure exactly what the problem is – and that's what you will do if you don't state "who is not doing what" – then your readers can't respond.

Three former consultants, Brian Fugere, Chelsea Hardaway, and Jon Warshawsky, address the fear of conflict head-on in their book devoted to corporate writing pitfalls. In a chapter entitled, "Kick the Happy-Messenger Habit," they exhort corporate writers to "stop tossing the rose petals."[2] Findings are often fertile ground for fallen rose petals, from what I've seen. Many of them devote at least a third of the text or space to what a team is doing well, or how hard staff members have worked, in order to soften the blow of the message that their efforts are in vain. This may briefly appease some readers, but it can dilute the overall message, leaving the reader wondering if there even is a problem. I've even seen some findings that are not findings at all, but simply opportunities to soothe egos by "tossing the rose petals."

This tendency may be stronger in some sectors than others. Emma Smith, Global Security Director at Vodafone, sees a clear difference between reports in banking and those in telecommunications. The latter, she says, feature "more data and metrics as part of the communication; shorter, more direct communication." Emma knows that a clear message doesn't come from data and metrics alone. Short, clear, concise wording helps your readers make sense of it all.

One example, from compliance reporting, shows how even slight wordiness can obscure meaning. The finding stated, "There is a wide range of instances of non-compliance in this area." Given that the report covered highly regulated activities within financial services, I was keen to find out how wide the range of non-compliance was. How many different types of non-compliance were there? I asked the writer, who said the sentence was supposed to convey many instances of *one* type of non-compliance. When I pointed out that these are two different problems, the writer was confused. "A wide range" or "many" – what was the difference?

If you have a wide range of different instances of non-compliance, there must be a problem with the people carrying out these tasks. How else could they all get so many different things wrong, repeatedly? You'd need to look at recruitment, training, management – all the steps that choose the right people and equip them to do the work.

If, however, there is only one thing that the team gets wrong, repeatedly, look at that thing. The problem won't be the people – it will be the processes or systems used to perform that task. Using language clearly and concisely helps readers understand what the problem is, which is essential to resolving it.

When your findings tell readers what they should be aware of or correct, they communicate problems – bad news. However, you will also want to point out good things you have discovered during your review or investigation. When your findings combine good and bad, you must find a way to make this clear to your readers. You can do this by labeling them differently ("to correct" and "good practice," for example), or by reflecting the positive and negative groups of findings in your report structure. What you cannot do is create a reporting layer-cake of good/bad/good/bad/good/bad and expect your readers to follow you.

"But the context!" I hear you cry. "We must include context!" Well, what do you mean by "context"? If you mean lots of supporting detail, revisit Chapters 5 and 6: how much of it is necessary? Can you put what is necessary in an appendix (the attic of your report)? Or must it go in the finding itself, risking making it longer and boring the reader?

If, however, by "context," you mean what I would bluntly call "excuses," then think carefully again about the purpose of the report. It is not a mouthpiece for departments or areas to explain – often without persuading – why they have failed. It is not an opportunity to distract from essential information with true but irrelevant details. It is there to document or prompt needed action.

This is the angle I've used when managers of audited areas have insisted I dilute findings with "context." If the "context" is the fact that an area is being restructured, or that two departments are merging, then that information belongs in both the terms of reference and the executive summary. It's essential for the reader to understand what is, in such cases, truly context.

However, in most cases, the conversation has gone like this (somewhat sharpened and simplified).

MANAGER: I think you've left out some important context here. You're reporting only what we haven't done – the governance framework.

ME: That's true. What do you think is missing?

MANAGER: Well, I already explained to you that we've been working on this for six months. I asked for extra resource, so now we have a team of ten on it, and we've even engaged external consultants to advise us. Don't you remember?

ME: No, I do. I remember you telling me all that.

MANAGER: So why isn't it in there?

ME: Mmm. Remind me, how much did you spend on the consultants?

MANAGER: Well, yes, we had to ask for extra budget for that...

ME: OK. Let me check I understand. You were supposed to have this in place last year, three months after starting. We're already three months late, and the budget exceeded by...I need to recalculate that to include the latest consultants' fees. You've also taken on extra staff from other departments. And you still don't have a framework in place. Now, I'm perfectly happy to put all of that in the finding for "context," if you insist. But do you really think it will focus readers' attention on the right things? Because, you know, I'm kind of worried it will make things difficult for you. I can just see different boards and committees demanding explanations, which only you can give, and that will take you away from your focus. Which is this framework. So what do you think?

[silence]

MANAGER: No, you're right, it's best to keep it simple. We don't want to clutter the report up with too much detail. I can get the framework in place by the end of the quarter.

Keep in mind that by focusing on "who is not doing what," you aren't blaming a person. You're defining a problem. "Define the problem and you're halfway to the solution."[3] The solution is ultimately to improve risk management and help the organization.

Jeffrey W. Brown, quoted in Chapter 6, sees how too much information, poorly articulated, can hinder this objective. Coming from an IT security perspective, he reminds his colleagues of the need to keep it simple and relevant for readers: "One of the things I like to drive in my team is clear communication. It doesn't matter how smart you are if you can't get your message through to other people. I remind my team to limit technical jargon and try to spend time understanding the businesses we are protecting."

Without understanding the business, you risk producing findings that waste time and improve nothing. However, where you work in the organization affects your findings. It's not only national culture (mentioned earlier), but also departmental culture that changes focus. Louise McKay, a former internal auditor, is a director of risk strategy. Whereas now she must report on trends and threats across an entire organization, "the challenge in internal audit reporting was to communicate detailed findings in a way that showed that I fully understood the area I'd reviewed – but also in a way that those who are less close to that area could understand the importance of what I'd found." The importance of what you've found starts with clearly stating "who isn't doing what" – this alerts readers to the fact that there is a problem. Why they should care about it, however, is the second Q: the risk statement.

SO WHAT?

This may seem a flippant question, given how serious the responses often are. But it's a good shortcut to articulate *why* your readers should care about your findings. They should care because, if they don't address the problems you've raised, the organization faces risks.

Now, different professions, sectors, organizations, and textbooks define the word "risk" differently. How could they not? Below is a selection of a few different definitions of the term.

- The Institute of Internal Auditors: "The positive or negative effect of uncertainty on objectives."[4]
- The Institute of Risk Management: "the combination of the probability of an event and its consequences. In all types of undertaking, there is the potential for events and consequences that constitute opportunities for benefit (upside) or threats to success (downside)."[5]

- PRINCE 2: "an uncertain event or set of events that, should it occur, will have an effect on the achievement of objectives. A risk is measured by a combination of the probability of a perceived threat or opportunity occurring and the magnitude of its impact on objectives."[6]
- ISO 31000-2009: "the effect of uncertainty on objectives"[7]
- UK Government: "specific uncertainties that arise from activities such as forecasting or implementation, the costs of which have been estimated. They are specific to an intervention and may be quantified and managed."[8]

There are some common points above – the emphasis on achieving objectives, for example. But it's clear that depending on context and audience, there can be wildly differing interpretations of the term "risk." If you and your readers are at cross-purposes when referring to risk, you're unlikely to get the results you need from the report.

Language matters. You need to discuss risk with your colleagues and clients so that you can agree on a common term. Many people use terms such as "impact" or "consequence," fearing that risk is too emotive. However, while something may be an impact, consequence, knock-on effect, repercussion, or fall-out from the problem you've observed, it's not necessarily a risk.

Look at your organization's risk-appetite statement. Decided by the board, it often features in the annual report. Align your work – whether in audit, risk, compliance, or IT – to the board's stated risk appetite or tolerance, and your reports will address the risks that most matter.

I've found it most useful to think of risks as falling into four main categories: financial, regulatory, reputational, and health and safety.[9] These all cover what we can agree are "real-world harms." When I worked in finance, we focused on the first three. The last one often applies to public-sector organizations in many countries (with their responsibility for law enforcement, prisons, and often health care) as well as manufacturing (with the risk of industrial accidents).

Some risk experts speak of opportunity risks, where an organization misses out on "the benefits of speculative opportunities."[10] For example, an overly cautious business may be reluctant to enter or expand its place in the market. Ultimately, though, opportunity risk maps to financial risk, as the business will not make as much profit as its less-cautious rivals. Similarly, existential risk – that of an organization ceasing to function – clearly has financial and reputational, as well as possible regulatory, risks.

For senior decision-makers and operational managers, financial, regulatory, reputational, health and safety, and possibly opportunity risks will capture their attention. This is because – unlike many, seemingly infinite, categories of sub-risks – these five hit the bottom line or the headlines. Without being too cynical, I believe readers best grasp risk statements that clearly state how the organization will lose money, receive bad press coverage, or even be held responsible for someone's life.

Most organizations have elaborate and well-thought-out risk categories, with numerous sub- and sub-sub-categories. This, they believe, helps them define specific risks departments face, and populate risk registers.

However, there is always a danger that by distancing themselves from the main categories, people fall into the trap of listing mere effects. Many of these will be failed controls – not risks, a point we'll examine shortly.

Let's look at two common types of risk cited in reports: cyber or data risk, and operational risk. In the first case, a common risk statement is to state the type of risk (cyber or data risk), then to elaborate on it, by saying something like, "Risk of data breach or loss."

Now, we all hope that anyone reading this will immediately understand why they should take steps to stop this from happening. But mapping this risk to one of the main risks – which can differ, depending on context – makes it much more tangible for readers.

Let's imagine a global telecommunications company – a mobile-phone giant, for instance. This company's audit, risk, compliance, or IT security team finds poor controls for customers' personal data. If there is a data breach (whether accidental or intentional), the risks are first regulatory. Where the EU's General Data Protection Regulation applies, this could entail a fine of €20 million or 4% of the organization's annual turnover, whichever is higher. There is clearly a financial impact, and, if reported in the press, reputational. Keeping in mind that GDPR protects EU nationals *wherever they are in the world*, this should focus minds.

Now let's imagine this same company's audit, risk, compliance, or IT security team has also found poor controls for sensitive proprietary data. This data is nothing to do with customers. It's what the company's research and development team has been working on to create a product that will surpass all rivals' offerings. Is the risk the same? Of course not. This time the risk is financial and reputational, in terms of missing an opportunity. The Institute of Risk Management definition above refers explicitly to the fact that risks can be both negative and positive – you can risk something bad happening (downside), or missing out on something good (upside). No company wants to see its chances of outpacing the competition and gaining greater market share escape. Yet that is exactly what report-writers need to communicate in this second example.

Emma Smith describes well the common failure to link observations to risk. "I often see technical people who understand their subject matter struggle to explain risks simply and clearly – to be able to share the 'So what?,' whether that relates to risks or generally the compelling reason."

Thinking of "So what?" in terms of the main risk categories set out earlier can also help avoid a mistake many report-writers make in their risk statements. Instead of stating a risk, they state a failed control – a different one from the initial observation ("Who is not doing what?"), but still a failed control.

One example that focuses people's minds on the difference is the following question: "If you're crossing a busy road, what is the risk?" Many people will answer "Not using the pedestrian crossing," or "Not crossing at the lights." Both are failed controls. The risk is that you'll be hit by a car and injured or worse. Crossing at the lights and using pedestrian crossings are controls to mitigate the risk of injury or death.

All functions that help organizations manage risk need to understand both risks and failed controls. However, when writing to a deadline, head full of detail, it's easy to confuse the two. For example, perhaps you've discovered that one department is using an outdated system for its own management information (MI). The system doesn't automatically capture the data needed, so staff members must enter it manually. This increases the chance of manual error, so managers need to check and double-check both the data entered and the outputs. What is the risk? Many reports will refer to the department's MI being late, or containing inaccurate data. Both are failed controls. If you ask, "So what?" you will immediately see this. So what if the MI is late? So what if it's inaccurate?

Many teams will simply say "operational risk" or "risk of operational inefficiencies." This may be true, but first, you must understand what the MI is *for*. If it is for internal decision-making about budget, then the risk is doubly financial. First, poor data can lead to poor decision-making, with the risk of financial loss. Second, "operational inefficiencies" simply mean wasted time and resources, which always costs the organization.

However, if the MI ends up in a regulatory report, the risks are different. You still have the wasted time and resources (financial loss), but now you must consider the regulatory risks. What happens if the regulator receives reports with wrong information? The organization could face regulatory censure, increased scrutiny (which in turn requires extra time and resource internally), fines, or even losing the license to operate in certain jurisdictions.

For compliance specialists, the "so what" may appear easy. The risk is of non-compliance with a regulation, law, or even organizational policy. However, there are ways to make this more useful for readers, who all too often see compliance as an internal, box-checking exercise, rather than an essential function.

First, look at the purpose of the regulation, law, or policy. It will be a directive control, telling people and organizations to do or not do something. If you can explain to readers why compliance is important – for instance, because it protects the organization from going bankrupt, or being sued – then they see the broader purpose. Furthermore, articulating risk in terms of real-world harms further persuades readers. Saying that there is a risk of regulatory censure is one thing; pointing out that a rival organization recently paid millions in fines for similar weaknesses is another. Keeping abreast of current events means you can paint a picture of what happens when organizations breach regulations and laws. Instead of generic phrases such as "increased regulatory scrutiny" or "significant fines," you can refer to tangible effects, such as newspaper headlines or dollar amounts.

The "so what" must *always* lead to something worthwhile for the readers – not to a list of seemingly trivial internal tasks. No matter how technical or complex the finding, you should be able to create a real-life risk scenario that is both proportionate and meaningful. Shawn Von Hagen of Canada Life refers to it explicitly, when articulating how his team bridges technical knowledge with organizational awareness to communicate with readers. Knowing what is relevant and keeping it simple are key:

> When writing technical audit reports, the team should at a minimum be familiar with COSO, COBIT, CIS and NIST. I always remind the team to know your audience and ensure that you have the *business so what* clearly articulated. When communicating a message, fewer words are better than more.[11]

Marisa Melliou echoes the importance of understanding the link between findings and objectives. "Remember to keep coming back to strategic objectives. Create value, business interest, and engagement by ensuring that audit objectives, audit programs, and report findings are always aligned to business objectives."

It's necessary, but not sufficient, to articulate "who is not doing what" and answer the question "So what?." For any report to communicate what can change, we have to include information that enables managers to correct problems. For that, we need the root cause.

HOW DID THIS HAPPEN?

This is where we discuss the cause of the observation. What happened to create the situation described in the finding, where people aren't doing what they should, or doing what they shouldn't? It's essential to discover this so that we can make sensible, sustainable, meaningful recommendations. Otherwise, the organization will never address what is causing problems. Existing problems will persist or recur, and new problems arise.

It seems obvious, yet many assurance functions fail to articulate any sort of true root cause of the problems they report. Whether in audit, risk, compliance, or IT, most reports either overlook root cause or simply restate the problem.

In November 2020, I spoke at an event for chief audit executives. The topic was root-cause analysis, and most of them complained bitterly that their teams were terrible at articulating root cause in findings. When I asked how many of their report templates specifically asked for root cause in the findings section, they suddenly went silent. The answer was that few of them had asked their teams for the information, but blamed them for not providing it!

Many people will say that they consider root cause as part of their investigations, research, or fieldwork. If this is true, then why isn't this crucial information making its way through to the reports? I think the answer is twofold. First, templates: if a template doesn't ask for information, few people will volunteer it. On the other hand, if a template asks for lots of other information, less relevant to the topic, people usually provide it without question. Badly designed templates have a lot to answer for, not least their role in giving report-writers an excuse for not thinking through what belongs in the report.

The second reason is that far too many people don't work all the way through to root cause. They may be satisfied with a circular or superficial explanation: "They're not following the process because, well, they're just not. They must be lazy." "The managers aren't reviewing their teams' work because they say they haven't had time." None of this is worthy of the term root cause. It fills space on the page, but doesn't fulfill any function. Come the follow-up, or the next review, you will find the same problems you found this time.

Root-cause analysis means persisting in peeling away layers of information to get to the true source of the problem. Some people use what they call "the five 'whys'," channeling their inner four-year-old. "Why did this happen? And why did that happen? And why? Why? Why?" It's demanding but effective. I've found that using the draft outline as an aide-mémoire (as suggested in Chapter 5) can prompt senior managers to start discussing root cause. Open-ended questions such as, "And why do you think that happened?," "What do you suspect caused that?," or even "Can you tell me how you think that came about?" often encourage people to share their theories and insights, all of which are useful in explaining how situations arise.

One of the most interesting results of examining root cause is discovering common themes or trends. For instance, in the following mind map, there are four findings, all with the same root cause.

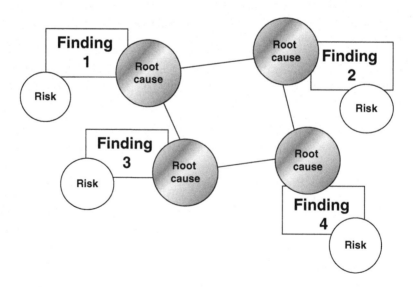

The obvious next step would be to see if you can combine all four findings into one. In this case, a team wasn't adhering to four different controls. The manager explained that the reason why was that he hadn't sent them for training. So on the face of it, lack of training is the shared root cause. However, combining all four findings into one gives a new, bigger finding: The team has not received training in key controls. After listing the risks, we come to root cause – *why* hadn't the manager sent the team on the training? Depending on the answer, you will probably have to ask more questions – and will definitely come up with different recommendations.

- The manager says he hasn't sent them on the training because the training is outdated and doesn't teach them what they need to know. This may be true – in which case, the training is an inadequate control. (If this is so, you should already have discovered this during your initial research. What's more, normally you wouldn't then proceed to test effectiveness. Why test how well an inadequate control works, after all? But I digress.) Why hasn't the manager put corrective controls in place, such as training people himself? Did he try to, but was stopped? If so, by whom? Why?
- The manager says he hasn't sent them on the training because turnover in the team is high, so it would be a waste of money. Why is the turnover high? Are there cultural or behavioral problems in the team? If so, what is the manager doing about it? Is he a problem? And why doesn't he understand the risks of *not* sending people on training?
- The manager says he hasn't sent them on the training because he can't afford to have even one person absent – there is too much work. Why? Is he managing resources badly? If so, why? Has he been overpromoted, or not received training himself? Or have his superiors given him too much to do without the proper resources? If so, why? Don't they understand the consequences?

You can see how even three scenarios can lead us into multiple lines of inquiry. Pursuing them, asking "the five 'whys'," peeling that onion, will eventually lead you to the true root cause, which is often cultural or behavioral. Even a list of seemingly minor findings – items "on the back burner," of low priority – may indicate a serious underlying problem: bullying, for instance, or incompetence at senior levels. Root-cause analysis can help you identify and resolve those problems before they lead to major incidents.

After all, items left on the back burner too long can still catch fire.

Without discovering and articulating root cause, findings will not lead to lasting action. Your reader should be able to follow a golden thread of logic and coherence from the observation ("Who is not doing what?"), through the risk ("So what?") to the root cause ("Why?") and ultimately your recommendation. If you skip root cause, your reader may well look at your recommendation and ask, "What exactly is the problem this is designed to solve?"

It's not simply about fixing a one-off problem, either. It's about resolving it for good, and reducing the chance of other problems springing from the same source. Imagine that a ceiling in your house is leaking. You need to find the source of that leak: is it from a damaged water pipe, or a missing tile on the roof? Once you find where the problem is, you stop that leak and prevent further ones. If, however, you simply spray an anti-damp product on the ceiling, or place a bucket under the leak, you've solved nothing. The problem will inevitably worsen. Findings without root-cause analysis are exactly that – saying, "Oh, there's a leak," placing a bucket under it, and walking away.

Another way to think of root cause is to imagine it as an illness, with findings as symptoms. If you go to the doctor because you have headaches, a tingling or painful feeling in a patch of skin, and a rash, you want him or her to tell you why. The diagnosis – shingles – is the root cause.

If you find yourself reviewing large strategic projects, using mind maps can help you discover bigger root causes than you may expect. A typical example will be where your findings include:

- no one signed off the final draft of the terms of reference;
- as a result, roles and responsibilities are unclear;
- reporting lines are consequently muddled;
- MI is contradictory and often repetitive;
- no one has received training; and
- communications are sporadic and poor.

If you list these on a Word document, it can be easy to see them as discrete failures or weaknesses in project management. But mind-mapping can sometimes open your eyes to the shared root cause – lack of governance – which is the major problem. The next step is to ask *why* there's no governance for such an important project. However technical a topic or activity, ultimately the root cause will be cultural or behavioral. Senior decision-makers either don't understand the risks involved – in which case, why are they in senior positions? – or understand them but don't care. Again, why would an organization promote such people? Especially in a technical context, these important points can disappear in a mass of data.

Jeffrey W. Brown sees this first hand in his work as a cybersecurity expert. "Security professionals share whatever information they can find or whatever is easy to compile. They provide way too much data and they don't provide enough business context around what the data means." Just as many auditors from accountancy backgrounds seek comfort in numbers, IT professionals rely on the data to speak for them. But, according to Brown, this overlooks what readers want and need. Most executives, he says,

> want answers to simple questions like are they secure and how do they compare with their industry peers. They want to know what they should be worried about, and they want to make sure their critical business processes are not impacted by a cyber event. Instead, they end up getting a lot of acronyms and technical jargon thrown at them. When you tailor your message to things that they care about, you can make a much better connection between security and the business. Understand your audience first and then tailor your message.

Finally, he says, "You'll need to make sure you're measuring the right things, not just the things that are easy to measure."

The security awareness professional cited in Chapter 6 echoes Brown's views while contrasting the approaches to technical findings. Speaking of analytical reports, this expert says that

> Often, reporting was heavily based on facts (events, people, dates, etc.), but limited "meaning" or foresight. This meant the reporting was akin to telling the news, rather than giving a decision-maker the possible foresight he or she required. Many analysts found "predicting" daunting and were reluctant to do so. Also, following the Chilcot Inquiry after the Iraq war, many analytical teams would stick to the facts in order to "stay safe" when writing. This potentially has led to a reduction in foresight being provided to decision-makers. It leaves the "sense-making" and deductions to them, rather than to the team providing the report.

How many reports can you recall that conveyed the right things to the right people in the right way? If they didn't, the problem probably started at the planning stage. However, our old companions – habit, tiredness, and fear – conspire against taking that step back, out of our comfort zone. Planning rigorously means that the testing, investigation, and research you do will be in the areas that most need it. Your findings will therefore be relevant and useful to your readers – but only if you convey them concisely.

It's often difficult to do so, especially if you've spent weeks or months working on a particular audit, review, or investigation. As we said in Chapter 5, planning techniques are essential for us to see clearly all the information in our heads, before we select and order it to best effect. One factor that more and more heads of department are admitting to is that people often don't complete their fieldwork or research before arriving at their conclusions. Since 2014, several heads of audit, risk, and compliance have told me that it is only when they see draft reports, and ask basic questions, that gaps in knowledge, testing, and analysis become apparent.

Oleksandr Tytkovskyi is Director of the Internal Audit Department of the Ministry of Defense of Ukraine – you can imagine how important his team's reports are. Like auditors the world over, he is alert to the causes of poor reporting. The first cause, he says, is low quality of the actual internal audit work itself. This ties in exactly with what the heads of audit were discovering late in the review process, as described above.

Poor-quality work, whether in audit, risk, compliance, or IT, can have many different causes. Tytkovskyi is open and comprehensive in his list of possible root causes:

> insufficient professional competence (lack of professional knowledge, skills and abilities); low motivation; propensity to commit acts of corruption; culture of poor behavior; adherence to outdated stereotypes; unclear and overly formal articulation of recommendations and proposals; and the inability to think creatively.

(I think it's inspiring that people from two fields often thought of as lacking in creativity – the military and IT – should stress its importance of creativity. Tytkovskyi here echoes Komitas Stepanyan's belief that we should think and act creatively, to have greater insight and communicate more effectively.)

This brings us back to the human element: creativity, but also culture and behavior. People may like to cite data and statistics to stay "safe," but without answering the three Qs, findings are meaningless. The only way you will be able to answer the three Qs is by asking them, or even more creative questions, and listening to the answers. Margaret McCaig is Global Head of Regulatory Compliance, Conduct and Legal Audit at a major international bank. She thinks one of the pitfalls of reporting is "too much background information, too much minutia." However, the wealth of detail doesn't reassure readers – report gradings can lead to resistance. "If red, there's pushback – sometimes even amber prompts it. Materiality of issues can also be an area that is often challenged."

Bombarding readers with data doesn't mean you'll persuade them. If you haven't stated clearly what the problem, risk, and root cause are, you're not serving your readers. If you leap to recommendations that are "quick fixes" senior managers will accept, you may be leaving massive gaps in the organization's risk and control framework.

Even specialists in IT security can lose sight of the "big picture." Professor Dan Shoemaker is Senior Research Scientist at the Center for Cyber Security and Intelligence Studies, University of Detroit Mercy, and 2021–2023 Distinguished Visitor at the IEEE Computer Society. He has an image few of us will forget when he talks about reports.

Far too many of the things I see are overfocused on the belly-button lint and miss the strategic point. There is a lot of designing of "solutions" in cybersecurity. These more or less protect some aspect without even considering any other (potentially easy) method of attack. The fact is that you really aren't secure if you aren't secure. So a report that concentrates on one of the three attack surfaces (for instance electronic) while ignoring the other two (human or physical) is a waste of time and money. My friends in the hacker community are looking for the gaps in the defense – not where your fortifications are.

Focusing on the three Qs – "Who is not doing what? So what? How did this happen?" – will prove that you indeed have a finding to report. Keep in mind, though, that a good report is one that conveys relevant, useful information. Resist the temptation to overstate findings, just as you should resist pressure to understate them. Everyone faces the "pushback" Margaret McCaig refers to; making sure that we have something that presents an actual risk helps persuade managers of the need to address it.

It's equally important to acknowledge when a failure or weakness isn't material. This shows clients, colleagues, and other parts of the organization that a finding isn't about "catching people out." It's about reporting what is truly relevant to the organization, to help it manage risks. Exaggerating or minimizing findings risks damaging your credibility and the organization's success.

This takes us back to points made in previous chapters. Clear, open communication helps at all stages leading up to the report, reduces conflict, and improves relationships. I've already mentioned using a draft outline as an aide-mémoire to draw other people out and encourage them to share information – having a "deskside" manner, much as a doctor has a bedside manner.

Maybe you are uncertain about this less formal approach, but professionals across the world have found that regular, often informal communication leads to better results. "We don't have a specific style of communication," says Ola Bello.

> The internal audit team is really close, and there are regular engagements and feedback within the team, so we naturally apply a similar approach to communicating with others without having something formalized. Our style is working collaboratively, and we have built good relationships over time with similar stakeholders.

Nino Karazanashvili introduced a different way of communicating into her team – one that prioritized regular, open discussions with senior managers across the organization. The senior managers, she said, saw this change and were surprised by the frequent meetings, and by our showing how we do our work, to make it more understandable for them. The internal audit team, as part of this process, saw the positive sides of being open in communication, which I think increased their motivation and responsibility. As for the management, it was a completely different approach to communication – even a bit "shocking." But it lowered negative attitudes towards internal audit.

With this approach, you can understand and incorporate senior managers' insights into your work as you go along. It's not about allowing them to dictate your work. Rather, it's about working with them collaboratively to produce the most accurate and relevant information. You may still have disagreements, but it should be easier to discuss and resolve them, building on the conversations you've already had. Many teams include a section entitled "management comments" or "management response" in individual findings. This is a good way to ensure senior managers have their say, without diluting or otherwise changing your findings.

Rather than doing all this hard work, though, maybe you'd rather hand it all over to an LLM, as mentioned in earlier chapters. As we have seen repeatedly, though, outsourcing our responsibility for thinking clearly and using plain language is not an option. Nor is overlooking what could be considerable legal and reputational risks if we use LLMs without controls over confidential data and the quality of the output. In a fascinating case study, Scott A. Emett, Marc Eulerich, Egemen Lipinksi, Nicolo Prien, and David A. Wood examined how one company's internal audit function used AI while controlling security risks. As Alee Marschke summarizes,

> To address security concerns, Uniper's information security department provided oversight of the audit, the company IP address was not used, and no personal or company data was used. The auditors did not provide full texts to ChatGPT. Instead, they used fragments of writing with ChatGPT and manually pieced the responses together.[12]

Writing a good finding is simple, but not easy. If you can't answer the three Qs, you either don't have a finding, or haven't finished your review. If senior managers disagree with your finding but cannot give any factual reason why, you will be on stronger ground if you've answered the three Qs. If they still disagree, you can invite them to add "management comments." But you will know you have findings that were worth discovering and are worth addressing.

As Emma Smith says, "Explain risk in real customer and business terms. Remove jargon. Remove or explain acronyms. Put yourself in their shoes. Use story telling. Data wins arguments. Be an enabler, not a blocker."

Now you're ready to write your executive summary.

ACTIVITY

Choose one of the case studies below and, with the limited (and sometimes confusing) information available, answer the following questions. Try not to write more than three sentences for each answer.

- Who is not doing what?
- So what?
- Why did this happen?

OPTION 1:

The process for selecting third-party suppliers is not formally documented. Management stated that as the division is a regulated function, all managers are aware of the importance of complying with agreed principles. Paperwork around suppliers chosen in the past provided to Audit showed consistency across the division, but there is no reference to Group policy. As it is a small function, staff members are told about processes in regular meetings (not minuted).

OPTION 2:

Data showed that 32% of staff members have IT privileges that enable them to access the sales database remotely (from a PC at home). Company policy is for managers to review access quarterly, even when there is high turnover. Approximately half of the managers do this, with the remaining 52% reviewing it annually (27%) or having no records of ever doing so (25%). This last group of managers could not confirm whether access for former staff had been removed, and there has been tracking of access from non-company laptops.

SUMMARY

- Any report will convey findings or observations, also called "issues," "key points," and even "weaknesses." The point of findings may be to document agreed action or to prompt recommended action. Structuring your findings to order information from broadest to most detailed will help meet these different readers' needs.
- One common approach in internal audit is the five Cs: criteria (or criterion, if singular), condition, consequence, cause, and corrective action.
- I suggest a simpler approach, which works for different types of reports, as well as business cases. Rather than the five Cs, it prompts the three Qs (questions):

- • "Who is not doing what?" (observation)
- • "So what?" (risk)
- • "How did this happen?" (root cause)
- This approach works well because if you cannot answer even one of the questions clearly and succinctly, you may not have a finding.
- When articulating risk, it is crucial not to fall into the temptation of confusing risks with failed controls.
- Root-cause analysis is essential for understanding underlying problems and making worthwhile recommendations.

ACTIVITY – SUGGESTED ANSWERS

OPTION 1:

- Who is not doing what? This division does not have a documented, approved process for selecting third-party suppliers.
- So what? The organization risks:
  - financial loss from poor value for money from suppliers who do not deliver quality products or services;
  - financial loss from legal action from unsuccessful suppliers, who may have grounds to sue if they discover that a non-existent process disadvantaged them;
  - reputational damage if suppliers have been chosen because of family or other connections;
  - regulatory censure in countries where private companies must comply with procurement law;
  - financial loss from inefficient use of resources (having to inform new and existing staff about the process verbally and informally instead of referring them to a single document).
- Why did this happen? It appears to be cultural – a small, tight-knit team specializing in a particular area may often feel they do not need to document what they do. Some of them may view the recommendation to do so as an insult to their knowledge and integrity. Others may simply have overlooked the need to document a process, believing that their existing arrangements were sufficient.

Original text with my commentary in bold:
The process for selecting third-party suppliers is not formally documented. **Note the passive voice.** Management stated that as the division is a regulated function, all managers are aware of the importance of complying with agreed principles. **This is a common excuse: "Because it's so important, we**

don't even need to document it – we live and breathe it!" This isn't good enough. Paperwork around suppliers chosen in the past provided to Audit showed consistency across the division, but there is no reference to Group policy. Did you spot the trap? Just because something is consistent doesn't mean it's good. It could be consistently bad. Also, have we checked that Group Policy's advice on this topic is consonant with current legal and regulatory requirements? It's no good having something in place that is compliant with a poor policy. As it is a small function, staff members are told about processes in regular meetings (not minuted). This is usually a sign a team is deliberately hiding information, or too casual about the need to document decisions and communications. Follow up.

ACTIVITY – SUGGESTED ANSWERS

OPTION 2:

- Who is not doing what? Managers are not reviewing staff IT privileges quarterly to restrict access to only those staff needing it.
- So what? The organization risks financial loss, since former – or existing – staff could disclose sensitive commercial information to competitors. If the information contains customers' or staff members' personal data, there is also regulatory risk. Both situations could also damage the organization's reputation.
- Why did this happen? It may be due to high turnover – we need more information. If the turnover affects only non-managers, managers may think it a waste of time to check access quarterly. If the turnover affects managers, too, then perhaps they have not received the necessary training.

Original text with my commentary in bold:

Data showed that 32% of staff members have IT privileges that enable them to access the sales database remotely (from a PC at home). **Ideally, we'd have more information here; it sounds like 32% of all staff, which could be the sales force. Or it could be 32% of staff in areas that don't need to access the sales database.** Company policy is for managers to review access quarterly, even when there is high turnover. **This is interesting – why** emphasize **"even when there is high turnover"? When I have seen this in fieldwork notes, it indicates the team or division in question has a history of using high turnover as an excuse not to perform crucial controls.** Approximately half of the managers do this, with the remaining 52% reviewing it annually (27%) or having no records of ever doing so (25%). This last group of managers could not confirm whether access for former staff had been removed, and there has been tracking of access from non-company laptops. **The syntax in this sentence is poor and therefore misleading. Does it mean the last group of managers couldn't confirm whether there has been tracking of access from non-company laptops? Or that they couldn't confirm whether access for former staff had been removed – but that there has indeed been tracking of access from non-company laptops? I think it's the former, as surely the latter would have included a statement about tracking results.**

NOTES

1 Richard Chambers, "Internal Auditors: It's What You Say – AND How You Say It!," "Chambers on Internal Audit," July 26, 2021, www.richardchambers.com/internal-auditors-its-what-you-say-and-how-you-say-it/; Norman Marks, "Let's talk about audit reporting," "Norman Marks on Governance, Risk Management, and Audit," July 26, 2021, https://normanmarks.wordpress.com/category/risk-2/

2 *Why Business People Speak Like Idiots: A Bullfighter's Guide* (New York: Free Press, 2005), 105.

3 Pressfield, *Nobody*, 33.

4 The Global Internal Audit Standards™ (Lake Mary, FL, 2024), 13, www.theiia.org/globalassets/site/standards/editable-versions/globalinternalauditstandards_2024january9_editable.pdf

5 *A Risk Management Standard*, 2002, 2, www.theirm.org/media/4709/arms_2002_irm.pdf

6 Prince2wiki, https://prince2.wiki/theme/risk/

7 International Standards Organization, 31000:2018, 2018, ISO 31000:2018(en), Risk management — Guidelines. See also UK Government, Orange Book: Management of Risk – Principles and Concepts (2020), 40, www.gov.uk/government/publications/orange-book

8 The Green Book, 2020. www.gov.uk/government/publications/the-green-book-appraisal-and-evaluation-in-central-governent/the-green-book-2020. See also UK Government, "Communicating Risk Guidance," 2011, https://assets.publishing.service.gov.uk/government/uploads/system/uploads/attachment_data/file/60907/communicating-risk-guidance.pdf

9 Chartered Institute of Internal Auditors, "Writing about risk" (technical guidance), reviewed 2020, www.iia.org.uk/resources/risk-management/writing-about-risk/

10 Paul Hopkin, *Fundamentals of Risk Management: Understanding, evaluating and implementing effective risk management*. 4th ed. (London: Kogan Page, 2017), 440. Hopkin's book, which is the study text for the Institute of Risk Management in the UK, sets out compliance, hazard, control, and opportunity risks.

11 See also Benjamin Power, "Writing Good Risk Statements," *ISACA Journal Archives*, May 1, 2014, www.isaca.org/resources/isaca-journal/past-issues/2014/writing-good-risk-statements and Ian Cooke, "IS Audit Basics: The Components of the IT Audit Report," *ISACA Journal*, January 1, 2020, www.isaca.org/resources/isaca-journal/issues/2020/volume-1/is-audit-basics-the-components-of-the-it-audit-report

12 Emmett, Scott A., Marc Eulerich, Egemen Lipinski, Nicolo Prien, and David A. Wood, "Leveraging ChatGPT for Enhancing the Internal Audit Process – A Real-World Example from a Large Multinational Company" (July 18, 2023). Available at SSRN: https://ssrn.com/abstract=4514238 or http://dx.doi.org/10.2139/ssrn.4514238 and Marschke, 9-10.

Chapter 8

Executive summaries

We come now to executive summaries – or, depending on your organization or team's approach, conclusions, key messages, opinion, or other terms. Whatever you call it, this part of the report encapsulates its message. It's similar to the headline, sub-heading, and first paragraph of a newspaper article, and serves a similar purpose.

Sawyer's Internal Auditing describes executive summaries in a way that applies to all reports – not only internal audit ones.

> Summaries come in many shapes and sizes. A summary may be a simple transmittal memorandum that sends the report to the president or other executives of the organization. Properly drafted, such a summary may be ideal for harried, harassed executives who want to read no more than they have to while reserving the right to dip into pertinent detail.[1]

For this reason, I'll use the term "executive summary," as it handily refers to both the most senior audience (executive) and the need for brevity (summary).

This chapter will explore the needs and expectations of those who rely on executive summaries. Who are your readers, and what do they want from executive summaries? It will suggest ways to meet those needs and expectations, as well as alert you to common misunderstandings and pitfalls when writing executive summaries. Finally, it will discuss common, often emotive, responses readers may have, and how to anticipate and manage them. You will also see real-life examples and a suggested outline, complete with prompts, to help you draft your report.

AUDIENCE AND PURPOSE

Who reads executive summaries, why, and in what context? The above quote from Sawyer indicates partial answers to these questions. The most senior people in your organization will read them, but so will many others. It's difficult to imagine why an operational manager, for example, would

DOI: 10.1201/9781003422365-11

skip straight to the detailed findings without reading the executive summary first.

Remember the point made in Chapter 6, which stated that most people rarely read beyond the executive summary. This brings us to the answer to "Why?": because the executive summary contains the most important information, without going into detail. It is akin to reading the first paragraph of a newspaper article – it should give you a good idea of the main message, but not force you to read the entire article unless you need or want to.

The one-page reports referred to in Chapter 6 should encourage excellent executive summaries. This, however, requires report-writers to step back from the detail of their findings and identify the common themes that run through them. So, if you find that poor governance or culture underpins what you have discovered, the executive summary is where you say so. You may expect executives, with their collective experience and intelligence, to infer from findings what the "big picture" is. However, even if that were true, it's not their job to do so – it's the report-writers'. Professor Dan Shoemaker reminds us that "most of organizational theory is oriented toward getting relevant information off the operational areas and up to the top." Report-writers best do this by delivering not only relevant information but also the conclusions drawn from it.

If newspaper article analogies don't work for you, consider a different one: a lawyer's opening statement to the court. In this instance, the lawyer will set out his or her position, supported by high-level evidence. "I will demonstrate to the court that my client is innocent of the charges, because he has a strong alibi, the forensic evidence is weak, and the witnesses against him are biased." In one sentence, you have the lawyer's executive summary, as it were. Cross-examination of expert witnesses – the detailed findings – comes later.

Ask any executive, or anyone who regularly reports direct to executives, and they will tell you the same thing: keep it simple, keep it short. I mentioned in Chapter 3 the CEO who said, "If you can't tell me what's happening in half a sheet of paper, then either you don't know, or you're trying to hide something." He had concluded that lengthy, wordy reports were a sign something was badly wrong in the team producing them – and that is often the case. However, even more often, people wrongly think that the more detail they give, the more they help the reader. As we saw in Chapter 6, structuring the report to include detail later in the report frees up the executive summary to do what it should: summarize for executives. Jane Bettany of Frontline AIDS says,

> Decision-makers want clarity – what is the report for, what is the decision to be made, and what are the options. I've seen very long reports where I get to the end and don't understand what the reason for the paper is, or what decision is required. Pitch the report at the right level with the right amount of detail. If you give them too much detail, they will interrogate it at that level and get lost in the mire.

Moreover, as she points out, not all executives or committee members share the same level of information about or involvement in the organization. Frontline AIDS is a charity, and the board members are voluntary trustees. As in any organization, some are experienced executives, others new to being on a board. "This means that we get a really independent view and fresh pair of eyes and can bring in expertise," says Bettany. "But it also means that pitching the information at the right level is again challenging. Executives need to have enough information to understand the organization and the question at hand, but not too much to get bogged down in operational detail. It's a tightrope walk."[2]

Readers of compliance reports are no different. "Senior executives want to understand if their risk and control environment is operating effectively, and, if not, how material are the deficiencies, and what are the potential consequences," says Margaret McCaig. "Some people want to justify their work by putting in a lot of detail on what they've tested and how they tested, but in my view, do senior executives really care? My view is that reports should be as short and easily digestible as possible."

You may think that IT or cybersecurity reports should allow more detail in the executive summary, since many readers are not technical experts. However, Connecticut State CISO Jeffrey W. Brown differs. Recall his point from Chapter 7:

> In the case of most executives, they want answers to simple questions such as, are they secure, and how do they compare with their industry peers. They want to know what they should be worried about and they want to make sure their critical business processes are not impacted by a cyber event. Instead, they end up getting a lot of acronyms and technical jargon thrown at them.

The key is to bridge the gap between the technical data and the business's needs. "When you tailor your message to things that they actually care about," Brown continues, "you can make a much better connection between security and the business. Understand your audience first and then tailor your message. The rest will come easily."[3]

Some organizations anticipate gaps between writers' and readers' understanding before reporting. Emma Smith at Vodafone describes how her teams work with board members to deepen understanding and set expectations. "Our board wants a tangible view of residual risk," she says.

> They want to know where the specific risks are and how we are working to reduce them. We use some training called the Cyber Series to make cyber real for our board and go through specific controls – the idea being they better understand how we manage the risk. They also visited our Cyber Defence Centre and spent a half day with the teams.

This may be the ideal, which not all organizations can achieve. However, understanding what executives need and want from reporting – which may be different things – is every report-writer's responsibility. Mark Carawan sets it out clearly when he says,

> I largely believe that what the board committees tend to want is what they need. I see the challenge being that there is a gap between what the board committees want and need, and what management wants to give the board and board committees. Management understandably is inclined to celebrate successes, defend positions taken, and elaborate on extenuating circumstances, mitigants and external factors that have frustrated internal strategies and tactics. Management needs to show constraint in overwhelming boards and board committees with what management wants to provide, and to focus on what the board wants and needs.

The next section will focus on how to meet your readers' needs through constructing concise yet powerful executive summaries.

MEETING YOUR AUDIENCE'S NEEDS: WHAT TO INCLUDE IN AN EXECUTIVE SUMMARY

Many people ask how long an executive summary should be. One rule of thumb is no more than 10% of the entire report. So a 20-page report could tolerate a two-page executive summary, in theory, and I have seen effective examples of this. However, my view is that people's attention spans are short – if you can keep the executive summary to one page, all the better. Every time someone turns a physical page of a document, it breaks the flow. A one-page summary – or an effective one-page report – is likelier to capture and keep the reader's attention.

Could your executive summary be even shorter? One chief audit executive (CAE) I knew would ask his team to email him final reports at attachments, with a two- or three-sentence summary in the body of the email. His point was that many senior readers received emails on handheld devices. If they could get the gist of the results without having to scroll down or open attachments, they were happy. It saved them time: they knew within seconds whether they had to download the report securely or could wait until later.

Once his team had provided two- to three-sentence summaries for each audit, the CAE would challenge them further. "If you can do this," he would say, "then why do I have to read long executive summaries?" The answer was that he didn't. However, we all know how easy it is to get into bad habits, such as including unnecessary detail in executive summaries. The CAE's team took the point and started producing executive summaries that really *were* executive summaries. This was a perfect example of the writers – internal audit – thinking of the readers, tailoring communications accordingly, and saving time. Ideally, a one-pager will do the same thing.

But what do you write in this miracle of conciseness? Just as I suggested three questions for findings ("Who is not doing what? So what? Why did this happen?"), I suggest two for executive summaries:

- What is the single most important message?
- How have you left matters with the client, business area, or senior management?

The first question appears simple until I challenge you to articulate it in a sentence. At this point, many of you will realize how often your "executive summaries" are merely condensed lists of findings. In other words, they repeat symptoms without giving a diagnosis. One head of internal audit proudly told me that he instructed his teams to copy and paste the detailed findings into the executive summary section of the template. They then whittled them down until they fit into three pages. Unsurprisingly, there was no single, overarching theme, message, or diagnosis. His team's executive summaries didn't do their job.

If you summarize your view of a client's or business area's control frame-work, the result will necessarily be high-level. I don't mean a boilerplate phrase such as, "Our opinion for this area/process [delete as appropriate] is: no/limited/reasonable/substantial [select one] assurance." Nor do I mean the color-coded rating many teams use.

What I mean is what you would say to the chief executive/head of the audit or risk committee/chief compliance officer/CISO [select one] if you found yourself in an elevator with him or her. Imagine that a senior person asks you for a quick update. It could be something like:

- "This area has worked hard, but still does not have an IT security framework in place."
- "This team relies on manual controls for 85% of processes, which creates a higher-than-acceptable risk of error."
- "The department has outsourced three of its five highest-risk clients to a third party without due diligence."
- "Staff members aren't aware of their responsibilities in avoiding data breaches through simple daily IT security controls."
- "Many senior managers have left, leaving gaps in knowledge and experience, with serious consequences for compliance."

You may at first think these look like findings because they are concrete. However, each one will have been based on multiple findings, all tending toward the same conclusion: a simple sentence such as those above.

It's not easy to do this – you have to make sure your findings are sound and that your root-cause analysis is in place. You also need a good dose of courage to deliver what is usually bad news. However, these elements, combined with the ABCs covered in Chapter 4, will impress your most senior decision-makers far more than any detail. They appreciate that you will have all the data, test results, observations, and other fieldwork evidence to hand in order to arrive at your conclusion. They'll appreciate even more the fact that you summarized it so concisely for them.

A simple executive summary requires detailed groundwork, but is exactly what readers need. According to Paul Breach of the London Stock Exchange Group,

> There is considerable focus on being concise. In fact I've heard the comment that board and senior committee papers need to read "like the front page of *The Sun*,"[4] by which people mean that there is an attention-grabbing headline and very little need to read many words thereafter to explain a given point. I think senior decision-makers want to get to the point quickly. A paragraph of context to get people "into the zone" at the start is probably fine, but the conclusions, in order of importance, need to arrive quickly thereafter.

If you meet the challenge of summarizing your conclusion in one sentence, you could add a few more sentences. However, the moment you get into detail that belongs in a finding, you've gone too far. Broader points that set the scene and allow readers to appreciate the importance of your conclusions are more useful. Mark Carawan says,

> I have always found it preferable to provide prior period comparatives, benchmarking from industries, geographies, and so forth to help give context and perspective. Where relevant, it is essential to state assurance specific to policy, regulation, and law. Problems tend to arise in determining whether you can reach a conclusion and provide reasonable assurance based on the work performed. Where there is a challenge in giving an opinion, then there is a tendency to ramble on in one's narrative, unfortunately not addressing what boards want and need.

This may seem like a lot of information, but with careful planning (Chapter 5) and plain language (Chapter 4), you can include it without losing the reader.

The second point that must be in an executive summary is how you left matters with the client, business area, or senior management. Often I phrase this as "next steps" when I discuss executive summaries during training, but I recognize this can lead people to think of recommendations or action plans.

What I mean is something much simpler. Has the client, business area, or senior management:

- agreed with your findings?
- agreed with them and – even better – already started addressing them?
- or, most important, refused to accept your findings but provided no evidence or reason why?

This last situation is one of the hardest for anyone in an assurance function to manage. By the time this happens, disagreements have probably erupted during fieldwork, and much more senior people than usual have started attending meetings. The atmosphere is already one of conflict, not cooperation. However, if this happens, *it must appear in the executive summary*.

Most teams will say that unless there is an explicit statement to the contrary, readers can assume cooperation and agreement from clients or business areas. This means, though, that on the rare and difficult occasions when they don't agree – and refuse to provide any evidence or rationale – the report must say so. Strangely, though, I've read perhaps half a dozen reports where the executive summary gave no indication of discord. Only when reading management's response to each detailed finding have I realized that not one has been accepted. In this case, the executive summary can feature a line or two saying, "Unfortunately, we were not able to agree our findings with X. We are of course always ready to receive any new evidence, and to discuss matters further." This makes it clear that there was a disagreement. The reference to "new evidence" should also make clear that those disputing the finding or findings did not provide a credible reason.

Oleksander Tytkovskyi of the Ukrainian Defense Ministry is extremely precise in what he includes in his reports. He emphasizes the need to comply with domestic professional standards and the code of ethics, maintain confidentiality of data, and prevent conflicts of interest. He includes in this list "the need to urgently inform management about obstruction or interference" in his team's work. When there is conflict, we must report it in order to resolve it – and hiding it in findings is *not* reporting it. "Never bury the bad news," says Mark Carawan. "Bad news must be within the first few lines of the first paragraph."

Burying bad news is common to most writers. After all, who wants to be associated with negative terms such as "weakness," "non-compliance," or even "failure"? The risks of doing so, however, mean that organizations cannot address problems. Worse, burying bad news can allow criminal activity such as fraud to continue, making the report-writer an accomplice. Consider the example in Chapter 1, where a risk team had more than enough evidence, but not enough courage, to say that senior managers were committing fraud.

Even in less severe situations, hiding bad news *is* bad news. It speaks to a culture where people feel they cannot speak out, or report what they have found. Maybe this is the case, as covered at the beginning of this book. Maybe it's perception – many report-writers assume their role is to "toss the rose petals," praise people, and put a positive spin on the negative. However, most senior decision-makers (in organizations with a healthy culture) *don't* want this. They want to know, as quickly and as clearly as possible, how things are. The news may be good, bad, or a combination of the two, but the writer's task is to convey this as effectively as possible. Komitas Stepanyan of the Central Bank of Armenia says,

> I think CEOs, boards, and audit or risk committees want is to get a real picture of the institution. Depending on what kind of person the report-writer is – risk-averse or risk-addicted – he or she may portray the same issue in a different light, which is real pity. Reports need to be balanced and unbiased, free of the subjective judgment of the report-writer.

We all use our judgment at work, but when judgment is clouded by fear, it fails. "Balanced and unbiased" means stating exactly how things are, without fear or favor. Part of achieving this is being strict not only about what you include in executive summaries, but also about what you exclude.

MEETING YOUR AUDIENCE'S NEEDS: WHAT TO EXCLUDE FROM AN EXECUTIVE SUMMARY

The two questions mentioned above – "What is the single most important message?" and "How have you left matters with the client, business area, or senior management?" – are non-negotiable. As with findings, if you can't answer those questions, you don't have a report.

Clearly, few teams will produce two-sentence executive summaries and leave it at that. However, how far can you go in adding material to the executive summary before you dilute your message and confuse or annoy readers? The conclusion I've come to, after reading thousands of reports, is: not very far. This is because once you start adding material, it's hard to stop. If you can justify including a paragraph about objectives and scope, you will likely be able to justify spending a page on them.

Ninety-nine percent of readers will skip that page.

You may have professional standards and national or local regulations that require certain information in the report. As we saw in Chapter 6, that doesn't necessarily mean it needs to appear in the executive summary – you may be able to include it in an appendix. Where you must include such information in the executive summary, challenge yourself to make it as concise as possible. So, for example, rather than copying and pasting entire paragraphs from your terms of reference, create a simple but meaningful sentence. This doesn't mean a vague, boilerplate sentence that is the same from report to report – again, readers will quickly learn to skip it. "The review's objective was to assess controls in [area] to ensure they were both adequate and effective" is generic. "This review examined whether procurement policies, processes, and systems in [area] comply with both the law and group policy" is more specific.

Useful advice comes from Philipp Kratzer, speaking of his time as Director of Galileo PRS Authority at the Austrian Federal Chancellery's Office for Information Security. His role overlapped central government, information security, and the European space program – all high-risk areas. His reports had to comply with numerous international and national standards and laws, as well as with his ministry's orders. However, Philipp's focus was always on the readers, who don't appreciate wading through references and acronyms: "Security is only as good as its weakest part. Security reports have to bear in mind that not everyone is a security expert, but they often use wording and abbreviations of specialized cyber/COMSEC/crypto/information security experts." Avoiding this pitfall is necessary to engaging non-specialist readers in the crucial task of improving IT security throughout an organization.

Many audit teams include a standard sentence in their executive summaries about why they performed an audit. Usually, this sentence says something such as, "We performed this audit because it was included in the annual audit plan, agreed by the audit committee." Or, for external or co-sourced audit teams, "We performed this audit as agreed in our contract with the client." In other words, "We did this because we planned and agreed to do this."

Such sentences don't just waste precious space on the page and everyone's time and effort – they train readers to skip them. Then, when you audit or review something exceptionally, that information will be lost. You may have to investigate an area following a scandal or other incident. I've certainly seen boards and audit committees request ad hoc work in such cases. However, if over dozens of reports you've trained your readers to skip the rationale for the work, don't be surprised when they do.

Most prefatory material in executive summaries is useless. Not once have I read a brief executive summary and wished for more. More often, I find myself wading through "executive summaries" whose sub-sections, entitled, "introduction," "background," and "context," have little to distinguish them, let alone justify their existence. And when there are too many sections and sub-sections, most report-writers will protect themselves by repeating the information in each. After all, that way, no one can accuse them of leaving information out! But think of your readers. By the time they get to page five, they are already bored and irritated, yet none the wiser about the single most important message.

One organization came up with a novel – to me, at least – solution: a "synthesis for committees," something I had never seen in thousands of reports. The explanation for this was that the executive summary was too long, so they produced a shorter one – a synthesis – for senior decision-makers. This overlooks the fact that the executive summary *is* for senior decision-makers – the term itself suggests this. However, the problem was greater than mere repetition of content. First, different people interpreted the readership differently. Some assumed the "committees" were the board, audit, and risk committees. Others assumed these were "committees" lower down the organizational ladder – operational steering committees, for instance. So, with uncertainty about both readers and content, it's not surprising there was duplication and confusion. There were even contradictory messages between the executive summary and the "synthesis for committees."

The longer the executive summary, the greater the contradictions. One draft audit report published by a UK local authority in 2009 featured an 18-paragraph, three-page executive summary.[5] Rather than subject you to the full document, I've summarized the "summary" below, with no attempts to clarify what was unclear in the original.

1. Rationale for the audit
2. Background about various committee meetings
3. Information about what various committees discussed during these meetings
4. Procedures before and at the time of the audit
5. A government report detailed elements of historic and current service to residents
6. More of the same
7. More of the same, with details of some promised fixes
8. A private-sector organization produced a review with proposals for future work
9. That same organization produced (another?) review in its research (review referred to in 8?)
10. More detail from "the report" (unsure which one)
11. Another meeting
12. Things the private-sector organization had done in the two years preceding this audit
13. Three government bodies have shared information
14. Two "main findings and recommendations" [both are recommendations, and "Who is not doing what?" isn't clear in either]
15. Assurance and risk ratings
16. Thanks to staff for cooperation
17. Disclaimer of responsibility
18. Glossary and explanation of ratings are in the appendix

This kind of "executive summary" is nothing of the kind. It is, most likely, the result of report-writers getting bogged down in detail while trying not to upset anyone. While I sympathize with their position – it cannot be easy reviewing anything that involves multiple public- and private-sector organizations – the result helps no one. By trying to avoid conflict, such reports often only confuse matters.

Komitas Stepanyan is clear about his most senior readers' needs.

What they don't want to see – what I think no one wants to see – is something that creates confusion or asks "To be or not to be?" CEOs, boards, audit and risk committees don't like when it's unclear what's expected of them. Sometimes it comes from lots of technical, low-level details, an unclear picture, or not enough research and effort into the options available.

From almost the other side of the world, John Chesshire echoes Stepanyan's views, if slightly more strongly. "What do senior decision-makers hate?" he says. "Lengthy, text-heavy narrative that delivers little insight and that focuses on the minutiae rather than the strategic, bigger picture context. I think many want solutions, rather than just problems – things that will make their lives easier and less painful." If you think obscuring or diluting your single most important message makes anyone's life easier – think again. You can help readers find solutions only when you've clearly stated the problems.

It's useful at this point to compare executive summaries from two different reports and imagine readers' reactions. Both are from national audit or accountability offices, and both are about national defense. The first is from the UK and – as with the local government report referred to earlier – features a lengthy executive summary. The National Audit Office (NAO)'s "Ministry of Defence: The Equipment Plan 2020 to 2030" is 53 pages long, with ten pages devoted to the executive summary.[6] The summary features sub-headings such as "Key findings" and "The Department's approach to producing the Plan." Yet the website page devoted to this report (and from which you can download it) offers a much more concise summary, entitled "Report conclusions".[7]

> For the fourth successive year, the Equipment Plan remains unaffordable. However, the Department has still not established a reliable basis to assess the affordability of equipment projects, and its estimate of the funding shortfall in the 2020–2030 Plan is likely to understate the growing financial pressures that it faces. The Plan does not include the full costs of the capabilities that the Department is developing, it continues to make over-optimistic or inconsistent adjustments to reduce cost forecasts and is likely to have underestimated the risks across long-term equipment projects. In addition, the Department has not resolved weaknesses in its quality assurance of the Plan's affordability assessment. While the Department has made some improvements to its approach and the presentation of the Plan over the years, it has not fully addressed the inconsistencies which undermine the reliability and comparability of its assessment.
>
> The Department faces the fundamental problem that its ambition has far exceeded available resources. As a result, its short-term approach to financial management has led to increasing cost pressures, which have

restricted Top-Level Budgets from developing military capabilities in a way that will deliver value for money. The growing financial pressures have also created perverse incentives to include unrealistic savings, and to not invest in new equipment to address capability risks. The recent government announcement of additional defence funding, together with the forthcoming Integrated Review, provide opportunities for the Department to set out its priorities and develop a more balanced investment programme. The Department now needs to break the cycle of short-termism that has characterised its management of equipment expenditure and apply sound financial management principles to its assessment and management of the Equipment Plan.

In comparison, in 2018, the US Government Accountability Office (GAO) published a report on Department of Defense financial management.[8] Entitled "The Navy Needs to Improve Internal Control over Its Buildings," the report features a "Highlights" page, which functions as the executive summary. It features the rationale, the results, and the recommendations in three separate boxes. While several of the sentences are very long, and the report could have used a bullet-list format for the numbered items, it uses graphics to enliven the text and help many readers grasp crucial points immediately. The title page alone, however, is a model of clear communication – would you need to read further to grasp the principal message?

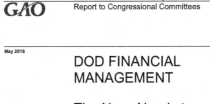

United States Government Accountability Office

GAO   Report to Congressional Committees

May 2018

DOD FINANCIAL MANAGEMENT

The Navy Needs to Improve Internal Control over Its Buildings

Accessible Version

GAO-18-289

Rachel Browne of Audit Scotland sees senior decision-makers wanting an executive summary (or equivalent) with clear recommendations for action in the report content and in any covering paper. She notes that Audit

Scotland calls its executive summaries "Key Messages, which may be statements or judgments (or potentially recommendations)." The NAO report uses the term "summary," while the GAO uses "highlights." Your team or function may use yet another term. What matters is that you articulate unambiguously the most important message at the front of the report. Your readers can then easily find, read, and respond to it.

EXECUTIVE SUMMARIES AND EMOTIONS

Putting the most important message upfront means, as Mark Carawan stated earlier, "Bad news must be within the first few lines of the first paragraph." Unsurprisingly, some people may respond badly to this. Even the most senior people have emotions. Astonishment, denial, anger, frustration, and defensiveness are all common responses to bad news.

Even if you have been at pains to report factually and carefully, giving quantified evidence in context, people may take it personally. After all, as we discussed at the beginning of this book, even the driest topic is important to people specializing in it. If you report that a team has not delivered a project successfully, in spite of spending time and effort on it, that team and its managers will be disappointed.

Using the outline technique described in Chapter 5 to structure regular discussions should help reduce the number of unpleasant surprises in a report. So too will including examples of good practice where you find them. However, it is not unusual for readers to focus on the one negative message even among a number of positive ones. It's human nature. Prepare for likely reactions by keeping readers informed throughout your audit, review, or investigation. If you discuss problems as they arise, you will build trust, which will also help readers approach the report more openly. "No surprises" should be the expectation by this point.

Most of you may be thinking of how corporate politics affects how we write and read reports. Yet corporate politics is not new; for as long as humans have cooperated, there have been factions, alliances, and power games in any setting. Keep it in mind, but don't let it dilute the value of your work. Knowing who is interested in your report and for what purpose is essential to planning a successful report. Include these people or teams on your mind map, so that you can, if necessary, speak to them ahead of time, or seek your own supervisor's advice about handling delicate relationships.

If you often see people "playing politics" with reports, that says something about your organization's culture. Are people routinely defensive, or keen to blame others? Are people afraid to speak out and therefore resist those who do? If the senior managers of the area you're reviewing behave in this way, you should consider addressing that in your work. It may not be easy, but it's certainly necessary.

As we said in Chapter 1, auditing corporate culture has been topical and indeed fashionable in recent years, with teams looking at this complex subject in specific functions or departments. However, whether you're in audit, risk, compliance, or IT security, culture is essential to your work. It will explain *why* people do what they do, whether good or bad. Without understanding this, you will never be able to give senior decision-makers the information they need to improve matters.

One CAE shared with me the wording used in executive summaries to introduce the topic of culture and behavior.

> During our audit, we observed behavioral aspects which by their nature are informal and less tangible. These behavioral aspects may be perceived differently by each individual, making them more difficult to review and to evaluate. These observations are opinions that can help explain why things go wrong; however, they are not subject to verification.

Unlike many audit functions, this team did not shy away from this crucial factor. The reports state upfront that yes, observations about behavior and therefore culture may be subjective – but they're important. After all, the observations can be positive as well as negative.

Balancing the positive and negative in your executive summary has two results. First, doing it proportionately means your message accurately reflects reality. Don't "toss the rose petals" or overemphasize the negative – either extreme will damage the report's effectiveness as well as your credibility.

Acknowledging what is good is not only the ethical and professional thing to do. It also reassures your readers that their successful efforts haven't been in vain. Rinus de Hooge urges report-writers to "report on what is done well, as too often reports contain only criticism, and not the positive messages. Try to step into the shoes of management and think about it how a report comes across if you were the recipient."

This isn't always easy, especially for those who are used to "reporting by exception." This approach says that if you don't mention something, your readers should assume all is well. The only reason you would mention something is if there's a problem. It's similar to the approach mentioned above, where report-writers expect readers to assume – unless stated otherwise – that clients, business areas, or senior managers have agreed to the findings.

While "reporting by exception" certainly keeps reports shorter and avoids ambiguity, it can also alienate readers and raise questions about the thoroughness of the report-writer's work. After all, if the report-writer tested ten areas and only one led to findings, where do we find information about the nine without findings? If any of them only just passed testing but are likely to fail next time, include this. If any of them are models of good practice for other parts of the organization, share the news.

Jane Bettany has been using the "appreciative inquiry" model in her work, but admits it doesn't necessarily come easily.[9] "I'm always trying to say something nice and be positive, which I find frustrating as I want to just say it as it is without faffing around." She also sees a risk in focusing on the positive, as readers themselves may overlook essential information, the "negative feedback when you have to improve."

Fran Meegan, an experienced quality assurance and training manager, always tries to be aware of what senior decision-makers may not be aware of. This isn't simply about "translating" technical terminology for non-specialists. It's about recognizing that what we assume is clear may not be to others. "They like to see things in reports that they have missed or would not have found themselves."

If you are in a similar position, consider the resources that can help you communicate complex topics – a common one is cybersecurity – to those who must understand them. Examples include the World Economic Forum's *Principles for Board Governance of Cyber Risk*, for instance, or the Conference of State Bank Supervisors' *Cybersecurity 101: A Resource Guide for Bank Executives*.[10]

Whatever your organization or function, certain principles will help your report reach its readers. Steven Sanders, Group Chief Internal Auditor at the Bank of Ireland, is adamant on the basics:

> Brevity, clarity, and precision! Any report content must be worthy of being read by a senior executive or board member. It's so important teams ask themselves the question, "Why would the board care about this?" If they can't answer that, then it shouldn't be in a report. No generic statements allowed! What exactly is the matter and what exactly is the impact? Don't be trying to talk to a senior executive or board member about general "risk of regulatory censure." Be specific. Be data-led. And be super tight on exactly what the problem is and what the first-order direct consequence is.

In one quote, we see the importance of knowing exactly what you have to communicate, to whom, and why. Without those elements, you have nothing to report. With them, you do – and must articulate them with "brevity, clarity, and precision." In other words, using plain language.

Whichever tools and techniques you use, plain language must be part of you and your everyday workings, mental, verbal, and written. As Alee Marschke says,

> We still have paper and pencil in addition to laptops. We still employ bank tellers despite the creation of the ATM. Our society may introduce

new tools and technologies, but we will not lose the old-fashioned trusted ways of doing things. AI tools will likely follow this same trajectory.[11]

Whether you draft by hand, direct onto a PC or laptop, or come to rely on LLMs for certain tasks, your writing is your product and your responsibility. Take pride in the thinking, the articulation, and the wording that convey your message.

And once you have done that, you are ready for the final – but often most complex – stage in report-writing: the review.

ACTIVITY

Use this outline to draft your next report. Try doing it by hand, and answer each question. Remember – if you cannot answer a question using plain language (active, brief, concrete), you should revisit your research or fieldwork.

REPORT OUTLINE

    I. Executive summary
        A. What is the key message?
        B. What is happening as a result of this report? (How has the client or senior management responded?)
    II. First finding
        A. Who is not doing what?
        B. What is the risk?
        C. What is the root cause? And your recommendation?
    III. Second finding
        A. Who is not doing what?
        B. What is the risk?
        C. What is the root cause? And your recommendation?
    IV. Third finding
        A. Who is not doing what?
        B. What is the risk?
        C. What is the root cause? And your recommendation?
    V. Fourth, fifth, sixth, and so on…
    VI. Appendix (Think of this as the attic of your report. Some readers may want detailed information about processes, regulatory guidelines, test results, etc. Put it here, rather than cluttering up the body of the report.)

SUMMARY

- The executive summary is the "headline" of the report. Whatever you call it – executive summary, summary, conclusions, key findings – your readers will look to it first. Therefore, the most important information – not details, but overarching messages – must feature.
- Knowing your audience and its purpose in reading the report will help you understand what to include. However, two non-negotiable questions are:
    - What is the single most important message?
    - How have you left matters with the client, business area, or senior management?
  (This is not about recommendations or action plans, but about whether the area reviewed has accepted your findings and, if applicable, ratings or gradings.)
- It is even more important to be clear about what to exclude. Lengthy prefatory or background material, detailed test results, and attempts to soften bad news with vague compliments risk confusing your message and distracting or annoying readers.
- Executive summaries, done well, may elicit strong responses. Managing relationships throughout your audit, review, or investigation reduces the chance of surprises and resulting conflicts.

NOTES

1 751.
2 For more information, see Sara I. James, "Effective communications - the report and the reporting process" (third in a series of articles communicating effectively with non-executive directors), *ACCA Internal Audit eBulletin*, November, 2020, http:// accaiabulletin.newsweaver.co.uk/accaiabulletin/1kfnflyqumr1g00djcpp5f?email =true&lang=en&a=1&p=58293902&t=28194286
3 See also Ian Cooke, "Your Audit Reports Have Consequences," *ISACA Now Blog*, February 12, 2020, www.isaca.org/resources/news-and-trends/ isaca-now-blog/2020/your-audit-reports-have-consequences
4 A UK tabloid and one of the most easily digested daily newspapers.
5 "Major Incident – Flooding. Final Audit Report," February, 2009, www.rbkc. gov.uk/pdf/flooding_auditreport09.pdf
6 January 12, 2021, 5–14, The Equipment Plan 2020–2030 (nao.org.uk).
7 www.nao.org.uk/report/equipment-plan-2020-2030/   "Report   conclusions" (277 words) follows "Background to the report" (237 words) and "Scope of the report" (84). This is typical of many reports, where the prefatory material – all the parts that readers skip, in Elmore Leonard's words – adds up to more than the substantive message. "Elmore Leonard's rules for writers," *The Guardian*, February 24, 2010, www.theguardian.com/books/2010/feb/24/ elmore-leonard-rules-for-writers

8 May 2018 www.gao.gov/products/gao-18-289.
9 Champlain University, "5-D Cycle of Appreciate Inquiry," https://appreciativeinquiry.champlain.edu/learn/appreciative-inquiry-introduction/5-d-cycle-appreciative-inquiry/
10 In collaboration with PwC, March 23, 2021, www.weforum.org/reports/principles-for-board-governance-of-cyber-risk; 2020 www.csbs.org/sites/default/files/2017-11/CSBS%20Cybersecurity%20101%20Resource%20Guide%20FINAL.pdf
11 Marschke, 12.

Chapter 9

Reviewing

Making the gain outweigh the pain

Most teams I have worked with over the past 15 years respond similarly when I mention report-reviewing. Sighs, frowns, rolled eyes, pointed glances toward certain colleagues – everything indicates that for most people, reviewing is the thief of all joy. But what do we mean by the term, and why do we put ourselves through the pain of the activity?

In theory, each review should have a specific purpose or focus, with the reviewer selected for his or her expertise in the relevant area. In practice, most people interpret review as anything from "mark up every single typo as well as questioning content and structure" to "completely rewrite the draft in my own style." It's no surprise then that people respond so negatively to the mere mention of reviewing. Yet it's necessary to produce something readable, credible, and therefore useful to the organization.

Often the review process takes far too long and improves little. Louise McKay of Royal London Group knows the stress it causes all too well.

> We all know colleagues who have been handed a vital report that's straying dangerously close to a deadline. This leaves them little time to review, never mind provide feedback that can be addressed before submitting the report. A clear structure and process, including rigid timelines, is essential.

When writers or reviewers are under pressure, they are likelier to make mistakes. Mistakes in a final report make it less comprehensible and credible.

But excessive delays, ostensibly to avoid mistakes, can leave people wondering if the report is even necessary. Richard Chambers, former President of the Global Institute of Internal Auditors, writes that he "once saw a report still being batted back and forth after a year."[1] I too have seen this happen, which leaves me wondering if there's any point in issuing the report. After all, if it contained important information, the time to act on it has long since passed. If it didn't, why report at all? Chambers is right – for some reports, it's better never than late.

Given how review can delay reports, this chapter will offer practical advice on how to review your own and others' work more efficiently and

DOI: 10.1201/9781003422365-12

effectively. It will define reviewing (as well as publishing terms for different types of reviewing) and suggest how to do so within teams. There is advice for writers and reviewers, as well as checklists for both.

REVIEWING, EDITING, AND PROOFREADING

We must define our terms at the outset. The best way to do so is by referring to how the publishing process works. If we consider standard usage and practice in publishing houses – whether trade, academic, reference, or any other genre – several useful distinctions emerge. (I'm excluding from this discussion *The New Yorker* magazine, which has a unique, multilayered, extremely rigorous editorial process,[2] and newspaper publishing.)

Good practice is to separate the editorial function from production. Editorial covers everything from accepting a manuscript to working on it for months, possibly years, until both author and editor agree it is ready to publish. The editor at this stage may be called a development editor, as he or she is responsible for helping the author develop the book's content. Structure, logic, tone – everything substantive is within the development editor's remit.

Once the author and the (development) editor agree that the manuscript is ready, it goes to the copyeditor. The *Chicago Manual of Style* splits copyediting into three sub-sections. The first, mechanical editing includes grammar, spelling, punctuation, and usage. This is what most people without publishing experience may call proofreading. The second, style, means adhering to a particular house or press style. That style guide will dictate everything from preferred spellings to bibliographical referencing. The third is substantive editing, which may include suggesting different ways to present the material.[3] This is not about duplicating the development editor's work, but about bringing a fresh perspective to the manuscript. A competent copyeditor should point out unnecessary repetition, clumsy or confusing wording, or illogical sequences, and can correct obvious mistakes.

However, the copyeditor *must not* rewrite the manuscript – if this is necessary, it is the author's job. One copyediting job I did several years ago specified in the contract that, in addition to putting the text into house style, my responsibility was "to ensure that the text is easy to read, grammatically correct, uses plain English where possible, and that the text isn't wordy with endless waffle." (And yes, I did suggest they remove the unnecessary phrase "with endless waffle"!)

Finally, proofreading is – in the publishing world – a different task from what most people imagine. My clients regularly talk about "wordsmithing" (always to accuse a reviewer of changing text to suit his or her own style) and "proofreading" (hunting typos and punctuation errors). True proofreading comes at the stage when a book is typeset – when it looks the way it will when printed. Benjamin Dreyer explains it as "a basic and mechanical process." His 30-year career in publishing started as a proofreader, and his "first jobs were simply to ensure that everything on the copyedited manuscript pages (you kept that stack to your left) had made its way properly onto the accompanying typeset pages (you kept that stack on the right)."[4]

My own experience of proofreading came in my early 20s, when I worked in a French government research institute. Long afternoons passed with indeed one stack on the left and one on the right. As I followed the text line by line on each page, side by side, I moved a ruler down each page so as not to miss anything. Proofreading – true proofreading – is, as Dreyer says, basic and mechanical, and also exhausting.

> Proofreading requires a good deal of attention and concentration, but it's all very binary, very yes/no: Something is right, or something is wrong, and if it's wrong you're expected to notice it and, by way of yet more scrawling, repair it. It's like working endlessly on one of those spot-the-difference picture puzzles in an especially satanic children's magazine.[5]

A crucial point is that these are not only different tasks but different departments. Editorial and production departments are separate for a reason – you need fresh sets of eyes at each stage. Even in a small publishing house, where staff may be experienced in different tasks, it would be risky for a development editor to also copyedit or proofread the same manuscript he or she has edited. You will all recognize how hard it is to check your own work for mistakes once you have read it multiple times. The same is true for those who edit and produce books. Word-blindness affects us all.

REVIEWING IN THE WORKPLACE

For those who write and review reports, I suggest not using terms such as "editing" and "proofreading." Instead, we should focus on reviewing substance and reviewing mechanics. Reviewing substance means exactly that – focusing exclusively on the message, and how the structure, logic, and tone convey it.

Reviewing mechanics is a completely different task: it's looking not at the substance, but the form. This includes spelling, grammar, punctuation, and usage, as well as house style and document layout. If your organization has a style guide setting out language preferences, how to write numerals and dates, capitalization conventions, and so on, you should follow it. Reviewing mechanics also includes checking page numbers, and whether running headers (the report title, for instance) are accurate and consistent.

It is entirely possible – although difficult – to review mechanics without engaging with the substance. It is essential to split the two tasks, though; trying to review both substance and mechanics will leave the reviewer feeling exhausted and irritable. What's more, he or she will have caught few substantive or mechanical errors.

These two types of review are different mental exercises, in the same way that sprinting and competing in a marathon are different types of running. You can't do both at the same time. What's more, most people will have a natural preference for one or the other. Just as professional sprinters rarely compete in marathons, so a reviewer with an eagle eye for mechanical errors may find it difficult to focus on substance alone.

I advise most teams to review mechanics only twice. When the writer is ready for the first review, he or she should check the draft report as thoroughly as possible: check spelling, grammar, and readability, and read the report out loud. If the writer has dyslexia, a review buddy can help at this stage. Then, after the report has been through however many layers of substantive review (I hope not too many!), someone who hasn't seen the report before should run a final mechanical check. Again, check spelling, grammar, and readability, and read it out loud. Many people think they can skip this second round of checking. However, if reviewers have tracked changes, including moving passages around, accepting those changes without checking the result can lead to what I call Frankensentences. You've all seen them – two sentences bolted together thanks to tracked changes, but with a necessary verb or transition accidentally deleted.

Apart from these two reviews for mechanics, the only reviews should be substantive. First, assuming the writer has thoroughly checked the draft report, there should be few if any mechanical errors to distract the reviewer. Second, if a reviewer is meant to look at substance but focuses on looking for typos, the report suffers. Not only does it miss out on the substantive review that should take place, but repeatedly reading a document looking for mechanical errors is a good way to become word-blind and to miss them. (There's also the fact that the reviewer may review a passage for mechanics and later decide to delete it altogether – a waste of time.)

REVIEWING YOUR OWN WORK

The first step to reviewing your own work is writing well. And writing well depends on everything we have covered previously: understanding yourself and your readers; planning; structure; and throughout, using plain language.

The reason should be obvious, yet most of us, bogged down in detail, miss it. The shorter and simpler your draft report, the easier it will be for anyone – you, your line manager, clients, other colleagues – to read it. As I said – obvious, yet easy to overlook. In previous chapters, I likened planning and structuring a report to building a house. Well, reviewing is like cleaning a house. It's easier to do if you've cleared the surfaces of all clutter. Even the novelist Colson Whitehead adopts this analogy: "Revise, revise, revise. I cannot stress this enough. Revision is when you do what you should have done the first time, but didn't. It's like washing the dishes two days later instead of right after you finish eating."[6]

Whether you ask professional writers or experienced corporate reviewers, the message is the same. Fran Meegan does what he advises when he says, "Keep it simple and succinct." If you do this, the review will be so much easier, not least because there will be less for would-be wordsmiths to seize upon. He sees writers creating a burden for themselves and reviewers by failing to heed his advice: "To impress reviewers and justify time spent, people fill up pages with unnecessary and irrelevant information." As we have seen in previous chapters, unnecessary and irrelevant information rarely informs and usually confuses.

Assuming you have followed the advice in the previous chapters, how can you help smooth the process? Again, advice from novelists can be surprisingly helpful. Zadie Smith says, "Try to read your own work as a stranger would read it, or even better, as an enemy would."[7] Although reviewers and other readers aren't our enemies, the point is valid. What would you notice if you were reading the draft with a critical eye? What weaknesses and gaps would you look for? Are you as alert to mechanical errors as the most pedantic reader?

Once you are certain your draft is the best possible, consider the reviewer's time and workload. Have you asked him or her to expect your draft? Even in teams where a workflow or work management system alerts people to tasks, it's rarely a bad idea to call or email people about an imminent draft. Even better, send a meeting invitation to discuss the draft once the reviewer has read it. Yes, you may send it, he or she may accept it, and the meeting may still be delayed. But the very fact of receiving the invitation will have planted a flag in the reviewer's mind: the draft report is coming on this day, and we should discuss it soon after. If the reviewer anticipates lots of tasks on a particular day, taking the time to alert him or her helps everyone plan their workload.

It's common sense and courtesy, but, submerged with tasks, we often unintentionally forget both. Remember, most people do not enjoy reviewing reports. It may be necessary, but it takes time, energy, and patience they don't always have. Steven Pressfield puts it bluntly but usefully when he says, "In the real world, no one is waiting to read what you've written. Sight unseen, they hate what you've written. Why? Because they might have to actually read it. Nobody wants to read anything."[8]

Another courtesy is to focus the reviewer's attention on particular tasks, or aspects of the draft. Most reviewers will not be aware of the difference between reviewing for substance and reviewing for mechanics. They will think they're being professional, conscientious, and thorough by trying to do both in the same review. Yet if you can persuade the reviewer to focus on one or the other, you will get a better result.

Even focusing solely on substance, most reviewers will give everything equal attention – until their energy or patience runs out. So specifying which sections of the draft would benefit from their insight will help them help you. It may be that the reviewer is a technical expert in a particular area; in this case, you can ask him or her to check you've explained it accurately yet simply, for non-experts. Or you could be concerned that your executive summary is too detailed, or sets out a conclusion that your findings don't clearly support. These kinds of comments help the reviewer focus on what may be a problem, and also set the scene for your discussion.

There should almost always be a discussion. Unless you have produced an exceptional draft for exceptional reviewers, who do not interfere unless absolutely necessary, there will be comments and changes.

Most people will say that they don't need a discussion. With most writing and reviewing done electronically, they rely on reviewers to track changes and insert comment boxes. Writers then usually accept most changes, except where they disagree. Then we see the dispiriting ping-pong of comment boxes in the margins as writer and reviewer argue their respective points.

Even when working remotely, a real discussion is possible. Whether by phone or videoconference, it is a more efficient and effective approach than the virtual ping-pong. Most people will say they don't have time for a half-hour or even an hour discussing the draft report. Yet they will spend much more time than that on the ping-pong, with worse results. First, the back-and-forth of the draft report with ever-increasing comment boxes takes time, too, and usually interrupts other tasks. Given that interruptions take several minutes to recover from, that time adds up. Second, "discussing" the report electronically removes the human element. If you *speak* to each other, whether by phone or videoconference, you will better grasp each other's views and feelings. This approach usually increases rather than diminishes understanding and goodwill. This can only benefit the current report, future reports, and the broader relationship.

As mentioned earlier, time – or at least a misconception of how best to use time – often gets in the way of effective review. People are trying to juggle multiple tasks, one eye on the clock, and often prioritize meeting deadlines over delivering quality. However, by thinking differently about how we use our time, we can deliver quality on time. It is – as with most worthwhile things – not easy, as it requires discipline and the ability to step back. If you have already managed to put at least some of the advice into this book into practice, though, you will have an advantage. You will already have experienced the benefits of forcing yourself to think and work differently, and seen how it improves the rigor and quality of your writing.

Take the lead in the reviewing process. By setting expectations about timing, focus, and medium, you can show the reviewer how to achieve more in less time. Then, when you in turn review other people's work, you can share with them the benefits of this approach. All you then need to do is discipline yourself further when reading the draft report.

REVIEWING ANOTHER'S WORK – AND HAVING
YOUR OWN WORK REVIEWED

The previous sentence is slightly tongue-in-cheek. Few things are harder than reading someone else's work and resisting the temptation to scribble all over it. You may be doing so from the best of intentions, but like an ill-trained puppy jumping up and putting its muddy paws all over someone, you'll only make a mess.

The preceding section is crucial to good reviewing. If the writer hasn't checked his or her work thoroughly – if you spot mechanical errors that spell- and grammar-checkers could have caught – then return the draft to the writer. Unless he or she has dyslexia – in which case a review buddy should have caught the errors – there's no excuse.

This immediately brings us back to the cry of, "But I don't have time! I need to review it now!" If you don't have time to do as I've just suggested, then three things will happen. First, you'll waste your own time on mechanical errors. Any substantive problems – the very things you *should* be looking for – will go unchecked. Second, you'll be doing the writer's work for him or her. Lazy colleagues love this, and will continue to rely on you to do what they should have done. Others, though, will never realize where they need to improve. Third, by reviewing quickly and badly *now*, you're guaranteeing that this inefficient and ineffective process continues, resulting in flawed reports.

Take that step back. What would happen if you returned the draft to the writer and negotiated a slightly later deadline for the report? The writer would realize where he or she has to improve, and would take responsibility for this and future reports. You, as a reviewer, would then receive a cleaner draft, which you can review substantively. The report recipients would receive a better-quality report – not just this time, but in the future.

When people say they don't have time to use this approach, they're saying quality doesn't matter. They're forgetting the purpose of reports and the readers' needs. Remember – no senior decision-maker ever said, "Ah, they've sent me the report by noon. It's completely unreadable, but at least they met the deadline."

If both writer and reviewer approach the process with common sense and courtesy, setting aside time to discuss their thoughts, the report will benefit. This requires both parties to change working habits, acquired over decades in some cases. It will probably also require the reviewer to abandon all assumptions about what reviewing is. It isn't, as we've seen, reading the draft report and trying to spot both substantive and mechanical problems at the same time.

Nor is it "wordsmithing," in the sense many people use the term. As stated in Chapter 4, "wordsmithing," which should be a fine and noble activity, is often used to mean endlessly carping and caviling over wording to no good end. You'll have met "wordsmiths" in the office; they're rarely fine crafters of elegant phrasing, spun from fine raw materials. As reviewers, they're the people who overlook major flaws in content to focus on mechanics. If the writing is wordy, passive, and convoluted, I sympathize. Faced with tens of pages of verbiage and a deadline, it's easier to focus on the superficial: typos (which the writer should have spotted), formatting, and imagined crimes against grammar.

Let's be clear – we're not talking about reviewers who spot mechanical errors and hand the draft back to the writer to resubmit a clean version. That's a good decision, and one that will improve the writer's skills and the team's efficiency. Nor are we talking about reviewers who are reviewing for mechanics: typos, formatting, and true grammatical errors are exactly what they should look for.

The "wordsmiths" are those who replace "however" with "nevertheless" – not for elegant variation, but throughout the document. They're the ones who rewrite in the passive, "because it sounds more professional." They leap on split infinitives and sentences ending with prepositions, circle them in red, and never realize that these "rules" are as contrived as they are unhelpful. (See Martin Cutts' chapter on grammar myths in *The Oxford Guide to Plain English* – it's an excellent defense against the "wordsmiths.")

Most damagingly, they're the ones who rewrite reports in their own style, following their own (mis)understanding of the topic. Without discussing the draft with the writer, errors can abound. After all, if a reviewer finds a passage unclear, by definition he or she is poorly equipped to rewrite it. It's all too likely the result will change the intended meaning.

A longer-term problem can arise from chronic rewriting. Years ago, a junior internal auditor in the bank where I worked told me that she'd received an email from her team leader. He told her that the first report she'd ever drafted had gone out in the final version. She looked it up and saw that – after five layers of review – she didn't recognize a word of her original draft in the final report. The content and style had both disappeared, replaced by five different reviewers' ideas of what she was trying to convey and how she should convey it.

Obviously, there is the risk mentioned earlier – that the message changes, and the report fails to communicate what it should. In this case, however, the review process had a further effect in completely demoralizing this young team member. No one had intended this to happen, but in the rush to meet the deadline, each reviewer had said, "Ah, I'll just change this to get it out, and discuss it with the team later." None of them did. Unsurprisingly, the junior auditor who'd drafted the report eventually left the function.

Many reviewers reading this will recognize the link between discussing reports with those who draft them and professional development. But again, time and habit get in the way. Rachel Browne of Audit Scotland describes perfectly the tension between what people know they should do, and what often happens under pressure.

It's seen as demotivating to rewrite people's draft reports – we try to balance getting the wording the way the reviewer wants it vs. using people's work as much as we can. We have high hopes for using active voice, plain language, etc., but local reports get written in a hurry, and the individual manager or engagement lead gets the final say on how we report.

Depending on the office politics, an influential but counterproductive senior reviewer can do lots of damage. However, most mean well and don't realize when they're not helping. There's also an element of passing on reviewing practices, even while recognizing their flaws: "My first-ever manager crushed my spirit when reviewing my work, so I'm doing it to you." Finally, there's the eternal existence of ego – most people instinctively impose their own style on others. Maybe they genuinely think it's better; maybe it is. But it won't help the writer improve.

Reviewers must take that step back – from the document and from their personal preferences. It's difficult because it involves emotions on both sides, and therefore requires not only common sense and courtesy, but also compassion. We rarely talk about compassion in reporting, yet it's essential. As much as we may hate reviewing, we have to assume that most writers do their best and submit their drafts in good faith. If we find errors, it's rarely because the writers have been lazy, sloppy, or unprofessional (although some may be). Most people simply need the objective – and compassionate – help of a reviewer to help them see what they no longer can. Benjamin Dreyer says of editors, "You're attempting to burrow into the brains of your writers and do for, to, and with their prose what they themselves might have done for, to, and with it had they not already looked at each damn sentence 657 times."[9] As reviewers, you too are there to help the writer overcome word-blindness and produce the best version of what he or she has to say – not what you wish to impose.

Early in this book, we discussed the importance of understanding our colleagues and readers – their assumptions, values, and yes, emotions. Not acknowledging how people *feel* about their writing – whether as writer or reviewer – leads to a vicious cycle. When reviewers rewrite, writers feel defensive, oppressed, or insulted. They then see no reason to try their best, as whatever they write will be rewritten. So then, they produce poor-quality work, which leads reviewers to claim they were right all along to rewrite. And the cycle – ineffective, inefficient, demoralizing – continues, producing reports that are so much less readable than they could have been.

Endre Bihari has a useful checklist of advice for reviewers. (You'll find another checklist, and a suggested reviewing process, at the end of this chapter.)

- "Casual skimming and superficial commenting does violence to the review process.
- "There is a difference between criticizing and critiquing. Make sure you do only the second one.
- "Do not take yourself overly seriously – it is not your writing.
- "Do not jump to conclusions. Work your way through to the overall end.
- "Keep examining your own biases as you read."[10]

This should make sense to everyone, although it can be difficult to force yourself to do it. One tip when reviewing substance is to not have a pen. If you are reviewing on-screen, cover the keyboard. The effect here is to deprive the reviewer of the chance to amend or make notes to the draft. If you're reviewing for substance, you should read the whole document through to the end, as Bihari says. Having a pen or keyboard to hand means you are almost guaranteed to use it – to make notes, to start suggesting changes, or to correct typos. All of these may be useful, but not at this stage.

If your first instinct when reviewing substance is to pick up a pen or poise your fingers over the keyboard, you will not only review badly. You will also demoralize the writer if you are visible. Put yourself in the writer's shoes. Imagine you're in the corporate panopticon of an open-plan office and have just handed your draft report to the reviewer. Now imagine the first thing you see the reviewer do is pick up a pen. You will probably interpret that gesture as proving that the reviewer intends to change your report whether it needs it or not, which is discouraging.

When working remotely, we often send our comments not by pen on paper, but by tracked changes and comment boxes. Think of the last time you opened a reviewed document. If it was covered in tracked changes, your heart probably sank. I understand – it's a depressing sight, and one that often puts the writer immediately on the defensive.

This brings us back to compassion. The point isn't to excuse or allow mistakes in draft reports, but to see and discuss them in a way that encourages the writer and makes him or her want to improve. We can only do this as reviewers by showing compassion for writers, and by reining in our urge to impose ourselves on their work. "Criticizing someone's writing can be unnerving for them," says Martin Cutts. "So, before you intervene, always consider how you yourself would feel about it. Remember also that editing is far easier than creating the first draft, so always give due praise."[11]

Cutts also usefully distinguishes between high quality and uniformity (often referred to as "consistency," by writers and reviewers who simply want to know whose style to imitate unthinkingly, in the vain hope that it will reduce the number of corrections). Give up, advises Cutts: "It's fair for you to want a high standard of clarity and grammar. But no two writers will ever use exactly the same words or tone of voice, so trying to impose uniformity will be both demoralizing and futile."[12]

Managers who understand the importance of reviewing well create processes to encourage both writers and reviewers to work constructively together. This could mean addressing directly the principles of communication, as Paul Breach does.

> In my former role as Head of Professional Practices, NatWest Group, I added a distinct chapter to our methodology focused on communication. I did so as I felt that it was necessary to bring together key expectations and standards in a single place. Also, by giving it a standalone chapter in a principles- and standards-led methodology manual, I felt it made a point in itself that this was an important topic which merited its own principles and standards. Quality assurance reviews, hot and cold, included particular focus on the quality of written communication. In addition, we quality-controlled a sample of audit reports before the reports were released.

Checking the report quality is essential in any department, and the results may influence staff members' annual appraisals. However, this should always be balanced, so that both writers and reviewers are held equally responsible. It's not fair to tell people that writing a clear draft report is a crucial part of their job, while not holding reviewers to account. Reviewers too should have to prove they have contributed substantively and constructively to the process – not simply rewritten other people's work. Both parties should also be able to demonstrate their willingness to have constructive discussions, by phone or videoconference, about draft reports. Sending multiple versions with tracked changes back and forth isn't as effective – it's a series of alternating monologues, rather than a two-way conversation.

Staff members will need different levels of support and comment about their drafts – good managers understand this. Jane Bettany of Frontline AIDS says of giving feedback,

> I used to be much better at this, but in my current role I have less time for it. I will often be sent a report to review and add comments with tracked changes, then send it back. With a team I trust, that will suffice; I will trust them to make the changes as they see fit and then send it out. With other staff members, who need a bit more development, I would ask them to send it back to me for review. I may go over it with them on a call in a bit more detail so they fully understand. The report may go back and forth a few times. I try to plan and set a deadline for the report to be sent to me well in advance of when it is due to allow time for this.

Not everyone is as sensitive to team members' skills, confidence levels, and needs, of course, and Rinus de Hooge thinks people in the corporate world could better manage their and others' emotions. "In most companies," he tells me,

> empathy is a word that is not on the radar screen of most managers. Neither is the word kindness. It is not expensive to be kind to people and still be a profitable and effective company. In an environment where people can be genuinely themselves, feel at home, feel safe, production is higher, cohesion is better and people stay longer with the company.

So how can you produce better reports with less effort at the review stage? Focusing on the difference between reviewing for substance and for mechanics is essential, as is *talking* to your colleagues. Don't outsource the process to a document with tracked changes and expect anything other than brief, factual responses. For the important substantive points – for example, why does the tone change halfway through? I thought the executive summary sounded positive, but the findings scared me! How can we articulate the risk better, so the readers understand why they need to act? – you need conversation.

Below is a suggested flowchart for the review process. You can adapt it to your organization, function, or team, as you see fit. *Sawyer's Internal Auditing* also has a good checklist for reviewing reports, including reviewing mechanics ("proofreading" in the text).[13]

For reviewing reports with clients or senior managers whose area you have reviewed, investigated, or audited, Sawyer features an entire chapter devoted to the topic. However, consider my suggestions from Chapter 5 – using an outline as aide-mémoire throughout your fieldwork can promote open, useful discussions with these people. As always, preparation can save everyone involved in the report process time, energy, and negative emotion. And the result will benefit the whole organization.[14]

PROPOSED SUBMISSION AND REVIEW PROCESS FOR REPORTS

Writer:

- Manage the process of drafting, writing, and reviewing own work, using 50/20/30 (50% of time spent on planning report, 20% on writing, and 30% on the review stage).
- Set expectations with peer or line manager reviewers about timescales and what you want from the review (technical expert's view? sense-check on risk ratings? review of mechanics?).
- Ensure your draft is as clean as possible – no typos or formatting errors.

Team (peer review):

- Review each other's work promptly and constructively.
- If the writer asks for a review of substance, do not check mechanics; if the writer needs help checking mechanics, do not meddle with substance.
- No wordsmithing – your individual styles are your own. If a piece of writing is clear, leave it alone.

Manager:

- Enforce deadlines and quality standards. If you see mistakes that could and should have been picked up at an earlier stage, *return the draft to the writer*.
- Check for substance – is it absolutely clear what the problem is, why it matters, and what the business needs to do about it? Is the substance relevant as well as true? Does it pass the "So what?" test?
- Are there any recent factors (current events, internal politics) that need to be taken into account?

Head of department:

- Support managers in setting and reinforcing standards.
- Review reports at a high operational or strategic level (no proofreading!).
- Contact managers immediately to clarify any points or flag any problems.

WRITING CHECKLIST

For *any* piece of writing, ask yourself:

- Do I know who the audience is?
- Have I made it clear what the issues are and why anyone cares?
- What is the order of events that will be clearest to the reader?
- Have I purged the draft of unnecessary words? Have I tried to keep sentences to 20 words or fewer?
- Have I avoided passive voice and zombie nouns?
- Have I used the most specific word possible in each instance? Have I used the correct word (there/their/they're, infer/imply, etc.)? Have I *made up* words?
- Have I checked spelling, grammar, and punctuation?
- Is the formatting correct?
- Have I read it aloud to spot any run-on sentences, clumsy phrases, or cut-and-paste casualties?
- Is this the best possible account I can give of my team's work and opinion?
- Could I send this to the executive now? In fact, if my manager weren't here to check this, could I send it straight out?

REVIEWING CHECKLIST

Before starting to review someone else's work (for substance, not mechanical errors):

- Put your pen or pencil away, or cover your keyboard so you cannot use the keys.
- Make sure you know who the audience is, at what stage the draft is, and whether you're reviewing for substance or mechanics.

While reviewing someone else's work:

- If you spot a mistake that a spelling- or grammar-checking tool could have picked up, consider returning the draft to the writer and asking for a clean copy.
- Read through the entire draft to form an opinion of the overall message and tone. Once you have done this, it's safe to use a pen or pencil!
- If something is unclear, make notes in the margins for discussion, rather than trying to rewrite.

After reviewing someone else's work:

- If it's not possible to discuss face to face, try to do so by phone. If you must comment electronically, use comments (Review/New Comment) rather than tracking changes. (Track changes only for mechanical errors.)

ACTIVITY

Below are excerpts from real emails from a banking organization in 2008. Various financial services organizations, central banks, and regulators had realized that the London Interbank Overnight Rate (LIBOR) was susceptible to manipulation, and therefore fraud. This became a scandal in the sector in 2012, leading to criminal charges in some cases.

Read the text below and see if you can review it first for mechanics and then for substance. You will find a list of mechanical errors on the next page.

Thank you for taking an interest in these matters and I hope that I have covered your concerns in saying that we will now consult both on a second and later Dollar LIBOR fix and on the transparency/stigmisation issue. Other clarifications are that we are strenghtening the governance immediately, that we have clarified the LIBOR definitions and that although we received no new application from any bank for any Panel, we will now open discussions with a selection of US and European banks for the purposes of seeking additional contributions particularly to the Dollar LIBOR panel. All other alterations are a mixture of better grammer, better layout and better explanations – the usual drafting points.

As to the prose, some we have used, some are no longer relevant as drafting has moved on, and some we will stay with our original phraseology.

Turning to the points of substance, practionners consider it a high risk move to say we will consult on the 12month fix. It will therefore be clarified in the text

Your proposal that we reduce what is said in sectiion 13 on governance I am afraid is not possible. In fact to get a way forward is going to require a stong governance for libor and the pressure is to do more not less.

I am also assuming that central banks will want to be associated with this in some form and look forward to the phraseology you want us to use.

The paper is now on its last round. Obviously we are not seeking textural and grammatical comments.

Now ask yourself:

- Which type of review came more naturally? Did you catch all the mechanical errors, or were you able to ignore them when reviewing for substance?
- How can you use this insight to offer reviewing services to colleagues?
- If you found all the mechanical errors, you are probably naturally good at this task. Volunteer to review a colleague's draft report for these – not substantive – problems. You may become the team's expert in this skill.
- Did your eye skip over typos and mispunctuation, while you focused on errors of logic or tone? Then your skill is in substantive review. This

means you can focus on content without being distracted by mechanical errors.

- It is good to know which your natural tendency is; mention this to your line manager, so that he or she can best use your skills to improve the team's reports.

EXERCISE – MECHANICAL ERRORS MARKED IN BOLD, WITH COMMENTARY FROM THE AUTHOR

Thank you for taking an interest in these matters and I hope that I have covered your concerns in saying that we will now consult both on a second and later Dollar LIBOR fix and on the transparency/**stigmisation [typo: stigmatization or stigmatisation]** issue. Other clarifications are that we are **strenghtening [typo: strengthening]** the governance immediately, that we have clarified the LIBOR definitions and that although we received no new application from any bank for any Panel, we will now open discussions with a selection of US and European banks for the purposes of seeking additional contributions particularly to the **Dollar [should be lower-case]** LIBOR panel. All other alterations are a mixture of better **grammer [typo: grammar]**, better layout and better explanations – the usual drafting points.

As to the **prose [odd choice of word – "wording" or "phrasing" would be better, as "prose" is more literary]**, some we have used, some **are** no longer relevant as drafting has moved on, and some we will stay with our original **phraseology [very odd choice of word – archaic and pretentious; "wording" or "phrasing" instead]**.

Turning to the points of substance, **practionners [typo: practitioners]** consider it a **high risk [should be hyphenated: high-risk]** move to say we will consult on the **12month [should be hyphenated: 12-month]** fix. It will therefore be clarified in the text

Your proposal that we reduce what is said in **sectiion [typo: section]** 13 on governance I am afraid is not possible. In fact to get a way forward is going to require a **stong [typo: strong]** governance for **libor [should be capitalized, as before]** and the pressure is to do more not less.

I am also assuming that central banks will want to be associated with this in some form and look forward to the **phraseology [see comment above]** you want us to use.

The paper is now on its last round. Obviously we are not seeking **textural [word choice: should be "textual"]** and grammatical comments.

Some may say that emails typed in haste are prone to more mistakes. While this is true, emails from other sources in the LIBOR e-correspondence featured few, if any mistakes. Furthermore, typing in haste would not explain the author's tendency to use overblown – and often inaccurate – vocabulary.

SUMMARY

- Reviewing is often among the most time-consuming and painful parts of the reporting process.
- Through understanding different types of review, and how and when to use them, writers and reviewers can work together to produce better final reports with less pain.
- Reviewing for mechanics should happen twice – at the very beginning and very end. It is about spotting typos, grammatical errors, punctuation and usage mistakes, and problems with house style or formatting.
- Reviewing for substance is about content – message, logic, and tone.
- Writers should do everything possible to provide reviewers with clear, concise, accurate draft reports, free of avoidable errors such as typos.
- Reviewers should focus on one type of review only, and avoid rewriting in their own style.
- Common sense, courtesy, compassion, and conversation – all are crucial.

NOTES

1 "For Some Audit Reports: It's Better Never Than Late," Richard F. Chambers & Associates blog, June 14, 2021, www.richardchambers.com/for-some-audit-reports-its-better-never-than-late/

2 The writer Mary Norris started her publishing career as "a page OK'er – a position that exists only at *The New Yorker*, where you query-proofread pieces and manage them, with the editor, the author, a fact checker, and a second proofreader, until they go to press." *Between You & Me: Confessions of a Comma Queen* (New York: W. W. Norton & Co., 2015), 12.

3 *The Chicago Manual of Style*. 14th ed. (Chicago: The University of Chicago Press, 1993), 65. See also Arthur Plotnik, *The Elements of Editing: A Modern Guide for Editors and Journalists* (New York: Macmillan, 1982), the Chartered Institute of Editing and Proofreading's "What is copyediting?," www.ciep.uk/about/faqs/what-is-copyediting/, and "FAQs: What is the

difference between copyediting and proofreading?," www.ciep.uk/about/faqs/ what-is-the-difference-between-copyediting-and-proofreading/

4 *Dreyer's English*, xxii.

5 xxii. In addition to proofreading and copyediting, I've also worked in the editorial department of a global publishing house, and as a production editor for an academic press. Looking back, I can see clearly that I was a good proofreader and production editor at the time, but did not become a good copyeditor until I was in my 40s. Part of this is down to the amount of leeway in each role – almost none in proofreading and production editing, but quite a bit in copyediting. With little leeway, the junior member of staff must restrain his or her impulses and simply follow the black-and-white guidelines. With time, and more manuscripts, comes a better sense of what is acceptable and what isn't. I strongly believe that poor copyediting usually comes from inexperienced copyeditors (often used because they are cheap). Even with a good style guide, copyeditors with little knowledge or experience may see ambiguity or error where there is none, and alter what they should leave alone.

6 "Colson Whitehead's Rules for Writing," *The New York Times*, July 29, 2012, www.nytimes.com/2012/07/29/books/review/colson-whiteheads-rules-for-writing.html

7 "Rules for Writers," *The Guardian*, February 22, 2010, www.theguardian.com/books/2010/feb/22/zadie-smith-rules-for-writers

8 Pressfield, *Nobody Wants to Read*, 4.

9 *Dreyer's English*, xxiv.

10 "On Writing and Reviewing…," EDPACS, 58:4 (2018): 1–28. 4, www.tandfonline.com/loi/uedp20

11 Cutts, *Oxford Guide*, 227.

12 *Ibid.*

13 787–91.

14 *Sawyer's Internal Auditing*, Chapter 14: "Audit Report Reviews and Replies," 819–48.

Conclusion

This book has taken you on a whirlwind tour of the most important aspects of report-writing. We began with the role of culture and cultures in communication and with reviewing draft reports. In between, we covered plain language and the importance of rigorous planning and structure, as well as common components of reports.

This book may be short, but I hope you feel it has been rich in advice and insights. Rather than focusing on theory, the point throughout has been to help you understand the underlying principles, motives, and habits that produce most reporting – so that you can do something different. Finally, words of encouragement – simply by reading even parts of this book, you have shown your commitment to improving your skills. This alone will make you a better professional, colleague, and communicator.

The very fact you have read this book, or at least parts of it, demonstrates that you know there are better ways to produce better reports. You recognize that "Reports are the auditor's opportunity to get management's undivided attention. That is how auditors should regard reporting – as an opportunity, not dreary drudgery – a perfect occasion to show management how auditors can help."[1] Whether you're an auditor, risk, compliance, or IT professional, reports can and should be exactly this – an opportunity for cooperation and improvement.

So what brought you to this book? Maybe you have finally realized that your organization's ways of communicating are ineffective. Maybe you have run out of patience with the way your colleagues or team members write and review. Or maybe you have decided to improve your own written work, reports included. Whatever the cause, you have taken a crucial step in your personal and professional development.

David Deegan is Executive Development Director (Special Operations) at Cranfield School of Management in the UK. He has worked with countless organizations and their staff over decades, building up insight into how and why people work the way they do. Using the "burning platform" analogy, he points out how difficult it is for people to shift their perceptions and actions.

DOI: 10.1201/9781003422365-13

Behavioral change does not happen without motivation. Essentially the leap into freezing seas from a 30-metre oilrig platform is dangerous and life-threatening – there is a possibility you will die. But the alternative is to stay on the oilrig, and if you do, your death is guaranteed. Although changing behavior is difficult, uncomfortable, and inconvenient, sometimes the impact of *not* changing is worse.

What are the risks your team or organization faces if it doesn't change its reporting? Often it will be the kind of risks mentioned in Chapter 7: financial, reputational, and regulatory. If severe enough, the risks could threaten the organization's very existence. However, if you work in the public sector, manufacturing, health care, or another field with physical risks, your organization could indeed face a real burning platform. Consider, after all, the BP Deepwater Horizon accident – I referred to the report several times in this book.

The point is that changing how we think, communicate, write, and review is one of the hardest things to do, wherever you work. It requires examining deep and long-held beliefs about ourselves, our place in the world, and how we relate to other people. Trying to do so in the face of corporatespeak, especially if it is ingrained at all levels, is even harder – all the more so as corporatespeak, for all its managerial trappings, is often fundamentally illogical.

As Barbara Ehrenreich says,

> I expected, as I approached the corporate world, to enter a brisk, logical, nonsense-free zone, almost like the military – or a disciplined, up-to-date military anyway – in its focus on concrete results. How else would companies survive fierce competition? But what I encountered was a culture riven with assumptions unrelated to those that underlie the fact- and logic-based worlds of, say, science and journalism – a culture addicted to untested habits, paralyzed by conformity, and shot through with magical thinking.[2]

The recent trend for "agile" auditing, loosely based on a method of software development popularized in the early 2000s, is an example. When one discipline lifts terms and principles from another discipline without fully understanding them, the result can at best be inefficiency. I have asked clients across the globe what they feel "agile" brings to their practice. The answers are often things they always could and should have done – regular meetings with clients and each other, for instance, or writing shorter reports. Often, though, they are doing the same things but with different labels, requiring a layer of less-than-intuitive vocabulary ("scrum," "stand up," "tribe") that people must learn simply to show they are team players. If this book does one thing, it should attune you to the language people use and why, so that you can decide for yourself what is useful – or merely a passing fad.

This book should do many more things, though. The first chapters focused on the importance of culture – individual, regional, national, professional, sectoral, and organizational – in communication. Organizations may not be people, but they may still have "personalities." When people talk about organizational culture, they talk about shared values and beliefs, but there is still room for different assumptions and approaches at both team and individual levels. Understanding ourselves and our colleagues and clients is essential to any successful communication.

Those to whom we communicate, though, can be outside our organizations. This broadens further the range of backgrounds, contexts, and objectives we must take into account when writing for a varied audience. If we don't understand our readers, who they are, and what they need, then our reports will fail the organization. And, as we saw through several case studies, the consequences can be severe.

Plain language can reduce the risk of reports failing to communicate well. Chapters 3 and 4 covered the theory and the practice of using clear, simple words and sentences. However, it requires clear-sightedness and courage to do so. Everything from individual ego to collective habit (corporatespeak) stands in the way, making writing clearly not only a difficult task but an ethical choice.

Writing clearly is as simple as the ABCs: being active, brief, and concrete. However, it's not easy. You need to understand your subject extremely well to be able to explain it simply. Favoring the active over the passive voice alone will improve any piece of writing. It will make it shorter, clearer, more confident, and compelling – all of which appeal to readers.

Before you start writing in wonderfully plain language, though, you must check you really do understand your subject and can explain it coherently. Planning techniques – mind-mapping and outlines, for example – will show you where you have gaps in your knowledge or research. Once you have all you need, they will help you organize your information in a coherent manner, including what is most relevant and useful.

The more you plan – sharpening your axe before chopping down a tree – the less you need to write. The James Ratio (50% of your report-writing time on planning, 20% on writing, and 30% on quality checks) works for everything from complex emails to exam questions to reports and books.

However, if your reader can't read your report, your efforts will have been in vain. Make sure the structure is simple and the layout easy on the eye. Everyone benefits from accessible writing, which is now mandated by law in many jurisdictions for public documents.

Reports require all the preceding elements to be in place: awareness of culture and its role in communication; understanding ourselves and our readers, including their purpose in reading reports; plain language, good planning, sound structure, and reader-friendly layout. If a single one of these is missing, the report risks being ineffective.

In the chapters on findings and executive summaries, I suggested questions that you must be able to answer in order to even have something to report.

- "Who is not doing what?" (observation)
- "So what?" (risk)
- "How did this happen?" (root cause)

and

- What is the single most important message?
- How have you left matters with the client, business area, or senior management?

If you cannot answer these questions clearly and succinctly, you are not ready to write. Have you understood risk correctly? Have you analyzed the root cause? Have you stepped back to see the overarching theme you must communicate? A report that simply lists stuff that happened may be true – but it is unlikely to be relevant or useful.

Once you have written your report, reviewing awaits. It can be constructive, productive, and build rather than damage relationships, but only if all parties return to the principles in the first two chapters. Understanding the people involved and their emotions, and learning how to rein in impulses such as rewriting or criticizing for the sake of it, are crucial.

Chapter 7 mentioned the five Cs approach to findings. I'd like to propose a different five Cs for reporting, or indeed any communication: culture, clarity, courage, compassion, and conversation. The last two appeared (with courtesy and common sense) in Chapter 9, with culture being the focus of the first two chapters. Clarity and courage are what we all need to see what is in front of us and name it.

Steven Sanders of Bank of Ireland sees only two crucial components of a good report: plain language and "guts." The first, because

> as soon as we humans – or auditors who happen to be human – pick up a pen, we somehow feel the need to be all officious and florid with our language. A career in politics awaits many an auditor with the lack of apparent desire to be clear. And that should not be seen as a good thing! I often read reports and have to read it again thinking, "What does this mean?" And guts because we can be so keen to not upset someone or take a clear stance that we sit on the fence. Jump off it and give a view. Newsflash – we live in a perpetual world of imperfect data. Have faith in your judgment and give a clear view.

Bringing together all the points covered in this book brings us to this place: one of plain language and guts. A profoundly human, reflective approach to report-writing does not come easily, requiring time, patience, and honesty. Understanding culture is a lifetime's work; seeing clearly means seeing what is bad and wrong; courage is prized precisely because it is rare; and compassion and conversation require more effort and discipline than most of us are used to. Yet without talking and listening to each other, we will never write anything worth reading.

NOTES

1 *Sawyer's Internal Auditing*, 729.
2 *Bait and Switch*, 226.4

Resources

According to Stephen King (yes, *that* Stephen King), "If you don't have time to read, you don't have the time (or the tools) to write. Simple as that."[1]

So if you want to write better, you must read more. I usually advise people to read as widely as possible, in any genre – except business and management books. (This book is an exception, of course.) The reason is that most books, whether fiction, popular science, political biography, even graphics novels, focus on communicating the writer's message to readers. This means that they show different ways of doing so.

Many business and management books, on the other hand, have as their sole purpose inducting readers into impenetrable and unnecessary jargon. If, however, your immediate purpose of improving your writing precludes forays into leisure reading, here are some useful books. I wanted to give my list of recommendations such headings as "Things I Like," but the excellent Benjamin Dreyer beat me to it. So I've organized them using slightly less personal headings: reference works, including style guides; language in the workplace; grammar and usage; history of the language; and writers on writing.

I'll start, though, with Orwell's six rules from "Politics and the English Language":

- "Never use a metaphor, simile or other figure of speech which you are used to seeing in print.
- "Never use a long word where a short one will do.
- "If it is possible to cut a word out, always cut it out.
- "Never use the passive where you can use the active.
- "Never use a foreign phrase, a scientific word or a jargon word if you can think of an everyday English equivalent.
- "Break any of these rules sooner than say anything outright barbarous."

You'll see he gives you a get-out clause in the last one – especially useful when you need to use the passive. If you need more advice, though, keep reading.

You may be surprised not to see some famous titles, organizations, or other language- and writing-related resources here.

This is not an oversight.

What you'll see below is what I've found most reliable and often laugh-out-loud funny. After all, if you don't enjoy dipping into a book or visiting a website, what's the use? Some of the items below will be familiar to you from previous chapters, while others will be waiting for you to discover them. In each category, I've included one or two "top picks" to start you off. Enjoy!

DICTIONARIES

- *Cambridge Dictionary*, https://dictionary.cambridge.org/
- *The Chambers Dictionary*, https://chambers.co.uk/book/the-chambers-dictionary/
- *Collins' Dictionary*, www.collinsdictionary.com/
- *Dictionary by Merriam-Webster*, www.merriam-webster.com/ (**top pick**)
- *The Oxford English Dictionary*, en.oxforddictionaries.com/
- *Random House Learner's Dictionary of American English*, www.wordreference.com/definition/

GRAMMAR AND USAGE, INCLUDING STYLE GUIDES (ALSO CALLED USAGE DICTIONARIES)

"I consider [a usage dictionary] kind of like a linguistic hard drive. To be honest, for me the big trio is a big dictionary, a usage dictionary, a thesaurus – only because I cannot retain and move nimbly around in enough of the language not to need these extra sources."

– David Foster Wallace[2]

- Articulate Marketing. *The Essential Business Grammar Guide*. www.articulatemarketing.com/the-essential-business-grammar-guide
- Brians, Paul. *Common Errors in English Usage*. Wilsonville, Oregon: Williams, James & Co., 2007. www.wsu.edu/~brians/errors/errors.html
- Bryson, Bill. *Troublesome Words*. London: Penguin, 2002.
- Capital Community College's Guide to Grammar and Writing. grammar.ccc.comnet.edu/grammar
- Cutts, Martin. *The Oxford Guide to Plain English*. Oxford: Oxford University Press, 2020. (**top pick**)

- Dreyer, Benjamin. *Dreyer's English: An Utterly Correct Guide to Clarity and Style*. London: Arrow Books, 2020.
- European Commission. "Claire's Clear Writing Tips." (particularly useful for non-native English-speakers writing in English) https://ec.europa.eu/info/sites/default/files/clear_writing_tips_en.pdf
- European Commission. "How to write clearly." https://op.europa.eu/en/publication-detail/-/publication/bb87884e-4cb6-4985-b796-70784ee181ce/language-en
- European Commission. *Style Guide*. https://ec.europa.eu/info/sites/default/files/styleguide_english_dgt_en.pdf
- Garner, Bryan A. *The Chicago Guide to Grammar, Usage, and Punctuation*. Chicago: University of Chicago Press, 2016.
- Garner, Bryan A. *Garner's Modern English Usage*. 4th edition. New York: Oxford University Press, 2016. (**top pick**)
- Fogarty, Mignon. *The Grammar Devotional: Daily Tips for Successful Writing from Grammar Girl*™. New York: St. Martin's Griffin, 2009.
- Fogarty, Mignon. *Grammar Girl's Quick and Dirty Tips for Better Writing*. New York: St. Martin's Griffin, 2008. www.quickanddirtytips.com/grammar-girl
- Fowler, H. W. *A Dictionary of Modern English Usage*. Oxford: Oxford University Press, 2002.
- Gowers, Sir Ernest. *The Complete Plain Words*. London: Penguin, 1984.
- Grammarist www.grammarist.com/
- Grammarphobia www.grammarphobia.com/
- Kamm, Oliver. *Accidence Will Happen: A Non-Pedantic Guide to English Usage*. London: Weidenfeld & Nicolson, 2015.
- Lynch, Jack, *The English Language: A User's Guide*. Newburyport, Massachusetts: Focus Publishing/R. Pullins Co., 2008 andromeda.rutgers.edu/~jlynch/Writing/
- Partridge, Eric. *Usage and Abusage: A Guide to Good English*. London: Penguin, 1963.
- Pinker, Steven. *The Sense of Style: The Thinking Person's Guide to Writing in the 21st Century*. New York: Penguin Books, 2014.
- Safire, William. *How Not to Write: The Essential Misrules of Grammar*. New York: W. W. Norton & Company, Inc., 2005.
- *SEC Plain Language Handbook* (1998): www.sec.gov/pdf/handbook.pdf
- Shertzer, Margaret. *The Elements of Grammar*. New York: Macmillan, 1986.
- Shrives, Craig, *Grammar Rules: Writing with Military Precision*. London: Kyle Books, 2011. www.grammar-monster.com
- Williams, Joseph M. *Style: Ten Lessons in Clarity & Grace*. 4th ed. New York: HarperCollins, 1994.

LANGUAGE IN THE WORKPLACE

- Collins, Philip. *To Be Clear: A Style Guide for Business Writing*. London: Quercus, 2021. (**top pick**)
- Fugere, Brian, Chelsea Hardaway, and Jon Warshawsky. *Why Business People Speak Like Idiots: A Bullfighter's Guide*. New York: Free Press, 2005.
- Garner, Bryan A. *Legal Writing in Plain English: A Text with Exercises*. Chicago: University of Chicago Press, 2001.
- Garner, Bryan A. "LawProse" blog www.lawprose.org/category/bryanagarner/
- Morris, Rupert. *The Right Way to Write: How to Write Effective Business Letters, Reports, Memos and E-mail*. London: Piatkus, 1999.
- Murphy, Elizabeth M. and Shelagh Snell. *Effective Writing: Plain English at Work*. London: Pitman Publishing, 1992.
- Phillips, Tim. *Talk Normal: Stop the Business Speak, Jargon and Waffle*. London: Kogan Page, 2011.
- Poole, Steven. *Who Touched Base in My Thought Shower?* London: Sceptre, 2013. (**top pick**)
- Seely, John. *The Oxford Guide to Effective Writing and Speaking*. Oxford: Oxford University Press, 2005.
- Simmons, John. *We, Me, Them & It: How to Write Powerfully for Business*. London: Cyan Communications Limited, 2006.
- UK Government Finance Function, *Good Practice Guide: Risk Reporting* (The Orange Book) (2021): https://assets.publishing.service.gov.uk/government/uploads/system/uploads/attachment_data/file/1010814/Good_Practice_Guide_Risk_Reporting_Final.pdf www.theiia.org/en/resources/knowledge-centers/artificial-intelligence/
- Van Emden, Joan and Jennifer Easteal. *Report Writing*. Maidenhead, UK: McGraw-Hill, 1987.
- Watson, Don. *Death Sentence: The Decay of Public Language*. London: Penguin, 2004.
- Watson, Don. *Dictionary of Weasel Words, Contemporary Cliches, Cant and Management Jargon*. London: Vintage, 2015.

HISTORY OF THE LANGUAGE, AND LINGUISTICS

- Crystal, David. *The Stories of English*. London: Penguin, 2004.
- Hitchings, Henry. *The Language Wars: A History of Proper English*. London: John Murray, 2011. (**top pick**)
- Greene, Lane. *Talk on the Wild Side: The Untameable Nature of Language*. Published under exclusive license from *The Economist*. London: Profile Books Ltd., 2018.

- Lynch, Jack. *The Lexicographer's Dilemma: The Evolution of "Proper" English, from Shakespeare to South Park*. New York: Walker Publishing Company, Inc., 2009.
- Pullum, Geoffrey K. *Lingua Franca* blog posts www.lel.ed.ac.uk/~g-pullum/linguafrancaposts.html
- Murphy, Lynne. "Separated by a Common Language" blog https://separatedbyacommonlanguage.blogspot.com/

WRITERS ON WRITING

- Garner, Bryan A. *Quack This Way: David Foster Wallace & Bryan A. Garner Talk Language and Writing*. Dallas: RosePen Books, 2013.
- Kilpatrick, James J. *Fine Print: Reflections on the Writing Art*. Kansas City, Missouri; Andrews & McMeel, 1994.
- King, Stephen. *On Writing: A Memoir of the Craft*. London: Hodder & Stoughton, 2000. (**top pick**)
- Moran, Joe. *First You Write a Sentence: The Elements of Reading, Writing...and Life*. London: Penguin Random House, 2018.
- Zinsser, William. *On Writing Well*. 25th-anniversary ed. New York: HarperCollins, 2001. (**top pick**)

NOTES

1 *On Writing: A Memoir of the Craft* (London: Hodder & Stoughton, 2000), 167.
2 Garner, *Quack This Way*, 69.

Index

Printed in the United States
by Baker & Taylor Publisher Services